MICROPROJECTION
WITH X-RAYS

MICROPROJECTION WITH X-RAYS

by

ONG SING POEN

TECHNOLOGICAL UNIVERSITY
DELFT

Springer-Science+Business Media, B.V.

1959

ISBN 978-94-017-6709-5   ISBN 978-94-017-6782-8 (eBook)
DOI 10.1007/978-94-017-6782-8

ERRATA.

page	line	instead of	read
9	18	$(1 - \delta)^3$	$(1/\delta)^3$
39	1	$\varphi = \approx 60°$	$\varphi \approx 60°$
39	fig. 6	fosusing	focusing
62	32	$\backsim_A$	σ_A
69	24	$\left(\dfrac{\varDelta D}{D}\right)$	$\left(\dfrac{\varDelta D}{D}\right)^2$
99	equation 25	$\dfrac{p^2\,e^a}{(\varDelta a)^2}$	$\dfrac{q^2\,e^a}{(\varDelta a)^2}$
100	6	μ^λ	μ_λ
107	14	leadacelate-	leadacetate-
110	fig. 12b	$\varDelta \gamma_r = \varphi\,\mathrm{tg}^2\,\gamma$	$\varDelta \gamma_r = \varphi\,\sin^2\,\gamma$

先嚴諱永立王府君鑒念

孝男盛本奉拜

to the

deceased

named

Ing

Liep

his   of the family of

son   Ong

Sing   respectfully

Poen   dedicated

humbly presenting

this work

CONTENTS

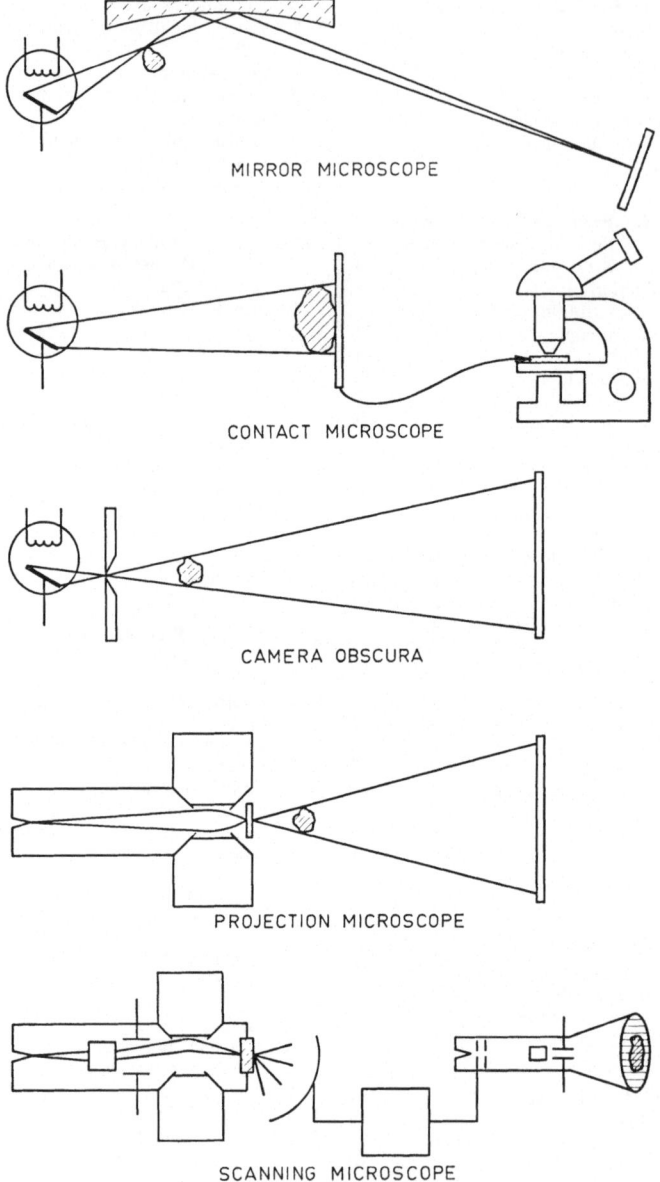

MIRROR MICROSCOPE

CONTACT MICROSCOPE

CAMERA OBSCURA

PROJECTION MICROSCOPE

SCANNING MICROSCOPE

The 5 X-ray microscope types.

CHAPTER I

GENERAL INTRODUCTION

§ 1. *Introduction.*

The desire to make enlarged images with X-rays is not new. Within two years of the discovery of these rays, Heycock and Neville[9]) published what they called micro-skiagraphs of alloys, and although with the magnifications used at that time one can hardly speak of microscopy, they laid the foundation for contact microradiography to be developed later on. The refraction of X-rays in matter was not perceptible with the apparatus available at that time, and in one of the first communications about his discovery W. C. Röntgen[19]) therefore concluded that lenses for X-rays could not exist.

Further investigations have proved that the refractive indices of all substances deviate only very little from unity. We therefore mostly write $n = 1 - \delta$, with δ of the order of 10^{-5} to 10^{-6}. As the focal length of a refracting surface is proportional to r/δ, r being the radius of curvature, it is almost impossible to make a strong lens. By using more components the rays are absorbed too strongly, and besides, such a lens would show large image errors. The spherical aberration constant of a single lens would be proportional to $(1-\delta)^3$. For light the case is much more favourable, a fact that is due to the existence of materials such as glass, having a high refractive index and very small absorption. For X-rays a high value of δ always coincides with high absorption.

In the course of years various types of X-ray microscopes have been developed that do not use lenses. Here follows a short description of each:

§ 2. *Mirror microscope.*

Although lenses for X-rays are impracticable, a useful image may be formed by means of concave mirrors. This is possible because the refractive index is smaller than 1, so that total reflection can occur. With the small value of δ the critical angle amounts to only a few minutes of arc. Total reflection therefore can only occur at grazing incidence. Thus a spherical mirror shows large image errors, whereas an ellipsoid has a very small image field. Baez and Kirkpatrick[10]) succeeded in 1948 in obtaining a resolving power of about 1 μ with a reasonable field of view. Another possibility is to make use of Bragg reflection which has been studied by Cauchois[3]) among others. The difficulties of this method are connected with lattice imperfections in the crystal.

§ 3. *Contact microscope.*

The contact method is, as mentioned, the oldest form of X-ray microscopy. This method consists of making X-ray pictures of an object in close contact with a fine grain film. The developed film is then viewed under the light microscope. The resolving power is limited by the resolving power of the film and that of the microscope. When using ultra fine grain film a resolving power almost equal to that of the light microscope can be obtained. Although with this method in principle any X-ray tube can be used, preference is given to a fine focus tube, where the current density in the focus can be increased considerably, since, according to Oosterkamp[16]), heat flow is more favourable with a smaller focus. Also, for a given depth of focus the distance between film and X-ray tube can be made small.

Very low intensities can be used as this method does not require focusing of the image. Thus the use of anode voltages of 500 V and less is no exception. Engström, Greulich, Henke and Lundberg [7]) have succeeded in obtaining a resolving power of 0.2 μ at these low voltages.

§ 4. *Camera obscura.*

The principle of the camera obscura dates back as early as the 16th century. Röntgen [19]) used this principle for proving the rectilinearity of X-rays. Using this pin-hole camera Czermak [5]) studied the emitting area's of X-ray tubes. For this purpose he recommended making stereographs. Sievert [20]) proposed the use of the pin-hole camera for microscopy in 1936, and Lutsau and Rovinsky [17]) reintroduced it at the conference on X-ray microscopy at Cambridge in 1956. Its operation is well known. It may be used for emission, as well as transmission microscopy. The resolving power is roughly 1 μ.

According to Engström[8]) this method offers the best prospects when using soft X-rays. As it combines the advantages of the contact method (no critical adjustment of focus required, large current density) and those of the projection method (the film does not limit the resolving power), Le Poole proposes to combine this method with the Delft focusing method (Chapter III). The size of focus can then be adapted to the maximum admissible current density. The difficulties in realizing this method lie mainly in the production of very small apertures.

§ 5. *Projection microscope.*

The idea of the projection microscope came from Von Ardenne[1]) in 1939. Independently, Marton[12]) and Cosslett[14]) studied this problem almost at the same time. Cosslett and Nixon[4]), however, were the first to realize the idea successfully in practice. In this type of microscope a magnified projection image of the object is made on a film or fluorescent screen with the aid of an ultrafine-focus X-ray tube. The magnification equals the ratio of the distances of the X-ray source to the film and that to the object. As the geometric blur of the image is a function of the magnification, intensity distribution and dimensions of the source only, the resolving power is, apart from diffraction phenomena, independent of the position of the object. Thus the depth of focus is very large. The resolution is approximately equal to the source diameter, which is minimized by using a strongly demagnifying electron optical system. The problems that appear in realizing such a tube form the main theme of this thesis. Up to now this type of microscope has given the best resolution, i.e. about 0.1 μ.

§ 6. *Scanning microscope.*

The idea of the scanning X-ray microscope is based on the micro-analyser of Castaing and Guinier[2]). In 1953 Pattee[15]) made an absorption X-ray microscope in which the object was in contact with a thin target, which was scanned from the other side by an electron beam. Cosslett and Duncumb[6]) have developed the Castaing method by giving a direct synthesis of the image, using a television tube in which the electron beam is synchronized with the scanning beam. Thus, immediate reproduction of the distribution of chemical elements in the object on the display tube is possible.

The scanning method has the advantage that magnification, field of view, and, in principle, contrast can be adjusted by electrical means. The resolving power is determined by the diameter of the electron beam at the object and the depth of penetration of the electrons. The depth of focus and the beam current are determined by the aperture of the electron lens, so a compromise should be found.

§ 7. *Other types.*

Beside these 5 types of microscope mentioned, there are other types, to be considered as modifications of the contact method. Instead of film, W. A. Ladd and M. W. Ladd[11]) use plastics for this purpose. Möllenstedt and Huang[13]) use a photo cathode, combined with the electron multiplier of Sternglass[21]).

As a modification of the projection method the microprojector of Avdeyenko, Lutsau and Rovinsky[18]) should be mentioned. Instead of the conventional target they use a pointed tungsten wire with a radius of curvature of about 0.1 μ. The advantage is that focusing is not so critical any more, resulting in a stable set up.

REFERENCES

1) Ardenne, M. von   Naturwissenschaften **27** (1939) 485.
2) Castaing, R., and   Proc. Conf. Elec. Micr. Delft (1949) 60.
   A. Guinier
3) Cauchois, Y.   Rev. Opt. **29** (1950) 151.
4) Cosslett, V. E. and   Natl. Bur. Stand. Symposium (1951).
   W. C. Nixon
5) Czermak, P.   Ann. Phys. **60** (1897) 760.
6) Duncumb, P. and   X-ray microscopy and microradiography, Academic
   V. E. Cosslett   Press inc., New York **(1957)** p. 374.
7) Engström, A.,   X-ray microscopy and microradiography, Academic
   R. C. Greulich,   Press inc., New York (1957) p. 218.
   B. L. Henke and
   B. Lundberg
8) Engström, A.   Personal communication.
9) Heycock, C. T. and   Trans. Chem. Soc. London **73** (1898) 714.
   F. H. Neville
10) Kirkpatrick, P. and   Jour. Opt. Soc. Amer. **38** (1948) 766.
    A. V. Baez
11) Ladd, W. A. and   X-ray microscopy and microradiography, Academic
    M. W. Ladd   Press inc., New York (1957) p. 383.
12) Marton, L.   Natl. Bur. Stand. Circ. 527 (1954) 265.
13) Möllenstedt, G. and   X-ray microscopy and microradiography, Academic
    L. Y. Huang   Press inc., New York (1957) p. 392.
14) Nixon, W. C. and   Lectures on the X-ray microscope presented at the
    A. V. Baez   University of Redlands (1956) 13.

15) Patee jr., H. H.   Journ. Opt. Soc. Amer. **43** (1953) 61.
16) Oosterkamp, W. J.   Phil. Res. Rep. **3** (1948) 49.,
                             Ibid, **3** (1948) 161., Ibid, **3** (1948) 303.
17) Rovinsky, B. M. and   X-ray microscopy and microradiography, Academic
    V. G. Lutsau   Press inc., New York (1957) p. 128.
18) Rovinsky, B. M.,   X-ray microscopy and microradiography, Academic
    V. G. Lutsau and   Press inc., New York (1957) 269.
    A. I. Avdeyenko
19) Röntgen, W. C.   Sitz. Med-Phys. Ges. Wurzburg (1895); Ann. Phys.
                             64 (1898) 6.
20) Sievert, R. M.   Acta Radiol. **17** (1936) 299.
21) Sternglass, E. J.   Rev. Sci. Inst. **26** (1955) 67.

CHAPTER II

PROPERTIES AND LIMITATIONS OF
THE PROJECTION X-RAY MICROSCOPE

§ 1. *Introduction.*

In 1951, shortly after the first publication of Cosslett and Nix-on[6]), Le Poole studied the possibilities of the projection microscope. Previously, he had acquainted himself with their difficulties, which concerned mainly the small intensity, and proposed the use of a stronger lens. Impressed by the results Cosslett and Nixon demonstrated later at St. Andrews, and convinced that with appropriate dimensioning a considerable improvement could be expected, Le Poole made an improvised projection X-ray microscope with old parts of the Delft electron microscope. The successful results from it made him decide to develop this type of microscope in Delft also, a work which unfortunately had to be stopped later on, due to lack of time. From 1954 on the author has continued this work under his supervision.

§ 2. *Short description of the projection microscope (fig. 1).*

The projection X-ray microscope consists in principle of a vacuum tube, containing an electron source, an electron optical system, and a target, which also acts as a vacuum seal. The electron lens forms a strongly demagnified image of the electron source on the target. The resulting X-rays emerge into the atmosphere, so that the object need not be placed in vacuum. It is situated just above the target, and the fluorescent screen, on which the image can be viewed through a 10 × magnifying eye piece, is placed further away. Efforts to develop the microscope as a sealed off tube have not succeeded up to now. Like the electron microscope the system is therefore made demountable, and is pumped continuously. For various reasons it is desirable that the target acts also as a vacuum seal.

a) The object can thus be put very close to the target, and

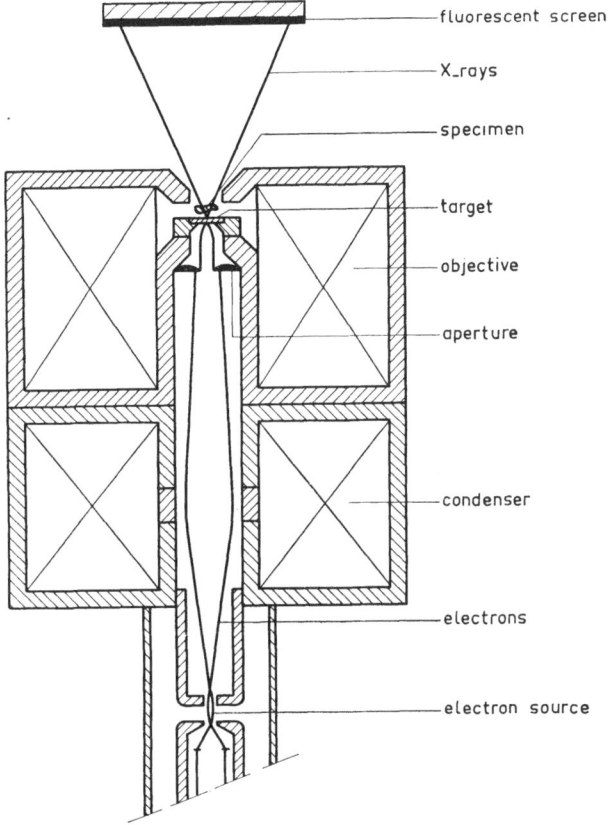

Fig. 1

Principle of the projection X-ray microscope.

still be in air. At a given magnification on the screen, the screen can come proportionally close to the X-ray source, which results in a brighter image or in a shorter exposure time.

b)   In connection with Fresnel diffraction phenomena it is necessary to place the object as close to the X-ray source as possible. See further § 4 of this chapter.

In general the lens system consists of two lenses, of which the first, the condenser, is weak, and controls the extent of demagnification. The other lens, the objective, is a strong lens, which finally forms the image on the target. The condenser is indispensable, which can be shown as follows: Due to image errors the focal spot cannot be made smaller than the diameter of the disk of confusion. An attempt at further reduction by a large increase of the

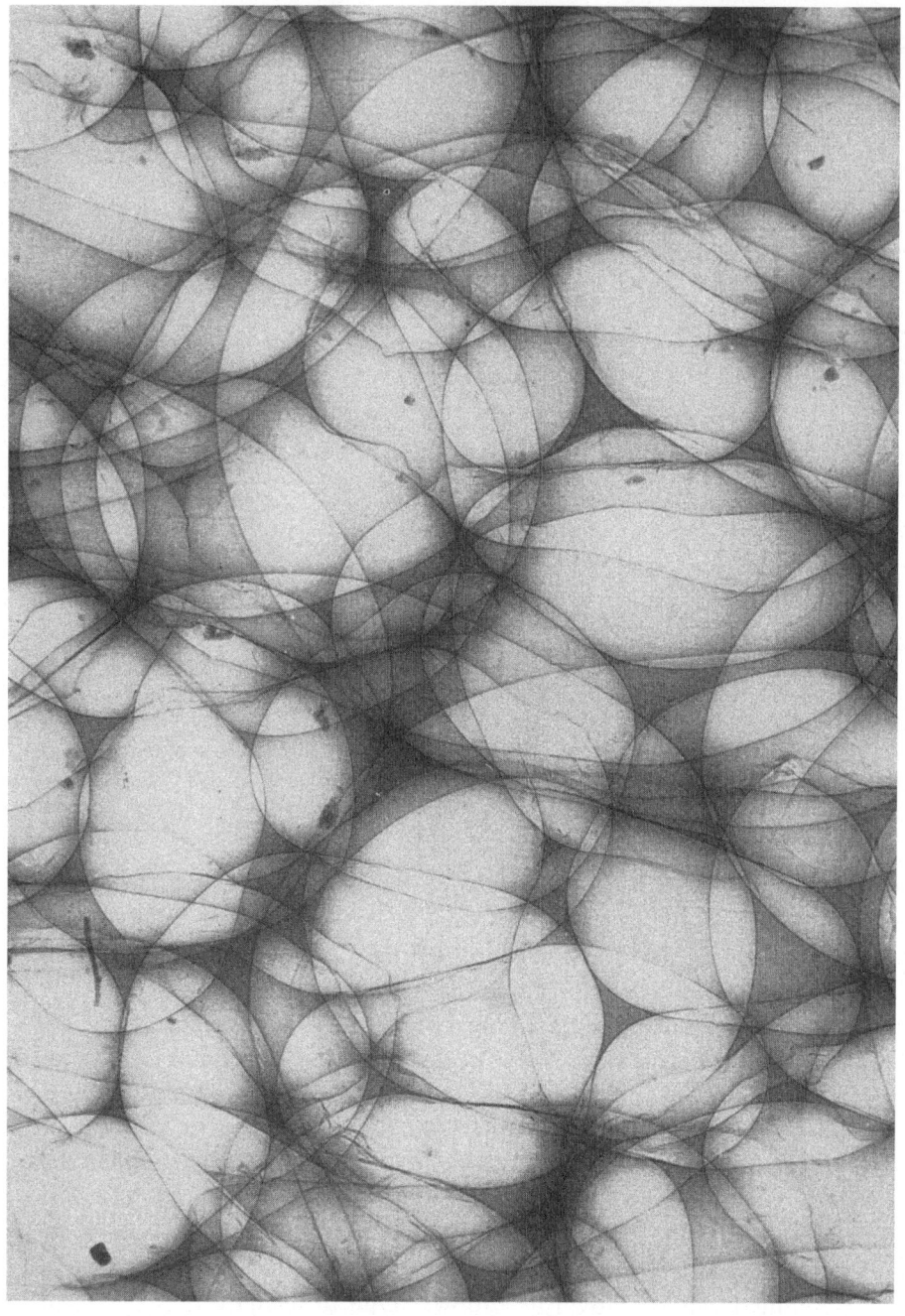

Fig. 2
Plastic sponge; Au target, 12 kV, magnification ca 90X. Due to the large depth
of focus combined with the great penetrating power of the radiation it is often
impossible to get an idea of the spatial distribution of the specimen.

demagnification of the system only results in a smaller current density in the focus. We must therefore adapt the demagnification to the image errors to get the optimum current density, which can be done only by means of the condenser.

§ 3. Properties.

The main features of a projection X-ray microscope are: large depth of focus and large penetrating power of the radiation. The first feature has been dealt with in the general introduction, the second one is clear without further explanation. These properties

Fig. 3

1500 mesh per inch silver grid, demonstrating the large depth of focus. The magnification varies widely over the different parts. Note the exactly correct perspective.

seem to be just what a microscopist may wish them to be. In practice, however, they give rise to complications in the interpretation of the image, for all parts of a three dimensional mass distribution will show equally sharp on a plane. Depending on the thickness of the object and its distance from the source, the various parts are represented at different magnifications. Due to the large penetrating power of the radiation the various parts are projected one on top of the other without giving indication on the spatial distribution. The result is that from the photograph we often cannot form a good notion of the object. As an example of such a photograph fig. 2 shows a plastic sponge. For such photographs of relatively thick objects we can hardly speak of "the magnification", as it varies for different parts of the object, possibly even by a factor of 10 or more (see fig. 3). As the image has perfectly correct perspective, stereoscopic pictures can be made approximating the ideal. The condition for making good stereographs is discussed in chapter VI.

The main disadvantage of the projection X-ray microscope is the low intensity, which makes visual observations almost impossible. This limitation also applies to the other types. In general the images are recorded photographically, the resulting photographs often being referred to as microradiographs.

§ 4. *Fresnel diffraction.*

With a real point source the resolving power of the microscope would be limited only by the occurrence of diffraction phenomena. The width of the first Fresnel fringe at an edge, calculated back to the object plane (see fig. 4a) amounts to

$$\delta_\lambda = \sqrt{\frac{\lambda a (b-a)}{b}} \tag{1}$$

in which λ is the wavelength, a and b are the distances of source to object and source to screen respectively. When $b \gg a$, which is almost always the case for the projection microscope, we can write for (1)

$$\delta_\lambda \approx \sqrt{a\lambda} \tag{2}$$

If the blur resulting from the finite dimensions of the X-ray source is larger than the width of the first Fresnel fringe, the latter can no longer be distinguished. For this reason Nixon[14]) suggested that

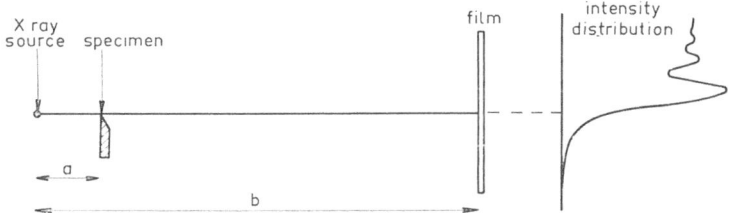

Fig. 4a
Resolution limitation due to Fresnel diffraction.

the size of the source diameter could be determined from the occurrence of the first Fresnel fringe and from the object distance a. Whether the fringe is perceptible or not however, also depends on the spectral distribution of the X-rays. As the fringe is not sharp in practice, and also as a result of the last mentioned cause, this method is very inaccurate, although it does give a good impression of the order of magnitude and of the astigmatism of the focus if any. Fundamentally, Fresnel diffraction only sets a limit when the resolving power is of the order of magnitude of the wavelength of the radiation used. If the diffraction plays an important part as a result of geometry (i.e. if the object distance is too large) an improvement can still be expected with the aid of Gabor's[11]) reconstruction method, where the attainable resolving power roughly

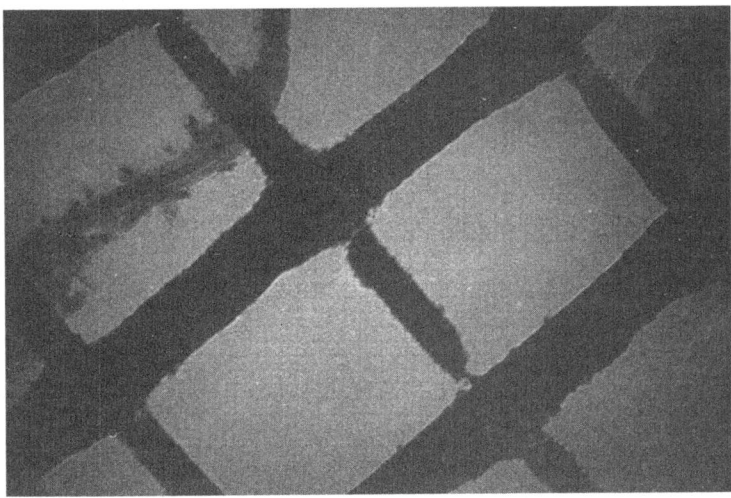

Fig. 4b
1500 mesh per inch silver grid. The first Fresnel fringe can be seen.

19

equals the distance of minimum to maximum of the last Fresnel fringe. Baez and El-Sum[3]) have examined reconstruction possibilities, but the results do not seem to be very encouraging, due to the polychromatic character of the radiation, and the large transmission of the object, which makes the occurrence of more than one maximum less probable. The first diffraction maximum can be observed with various objects (see fig. 4b). As yet it does not constitute a real limitation to the resolving power as the object distance can always be made small enough.

§ 5. *The finite size of the X-ray source.*

Of a much more serious nature is the limitation of the resolving power as a result of the finite dimensions of the X-ray source, whereby a blur (penumbra) appears in the image. Two particles cannot be distinguished if their penumbras overlap too much. If the source diameter is d, and the object- and film distances are

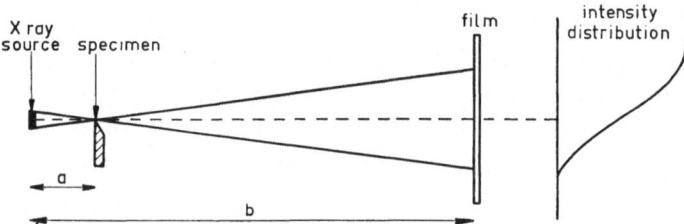

Fig. 5
Resolution limitation due to the finite size of the X-ray source.

respectively a and b, the width of the penumbra (see fig. 5) is

$$p = d \frac{b - a}{a} \tag{3}$$

The magnification on the film is

$$M = \frac{b}{a} \tag{4}$$

so that

$$p = d\,(M - 1) \tag{5}$$

The resolution as a result of the finite dimensions of the X-ray source is then

$$\delta \approx \frac{p}{M} \approx \frac{d\,(M - 1)}{M} \tag{6}$$

If $M > 1$, which is almost always the case here, then is

$$\delta \approx d \qquad (7)$$

§ 6.   *Contrast and details.*

As a microscope is made for observing small details, we must see to a sufficiently large contrast. On the basis of experiences in X-ray diagnostics, the thickness z of the object must satisfy the following equation to distinguish it from the surroundings.[1])

$$z = \frac{8 \cdot 10^{-4}}{\varrho \, \lambda^3} \approx \frac{10^{-7} \cdot V^3}{\varrho} \qquad (8)$$

ϱ is the density in g/cm^3, λ the wavelength in Å, V the accelerating voltage in kV, and z the thickness of the object in cm. Although this equation applies to white radiation at high voltages, it does give an impression of the order of magnitude. For biological objects, where $\varrho \approx 1$ g/cm^3, V must be $\approx 4{,}6$ kV ($\lambda \approx 4.3$ Å) to just distinguish an object of a 0.1 μ thickness from its surroundings. If we want to observe differential absorption a much lower voltage must be chosen. According to Engström, Greulich, Henke and Lundberg[10]) a voltage as low as 0.3 to 2 kV (60 to 10 Å) is necessary to obtain optimum contrast with biological objects of 2 to 6 μ. At the resulting extremely large wavelengths the object must be inserted in vacuo. These low voltages can only be applied in contact microscopy. In the projection microscope the anode voltage cannot be made substantially lower than about 4 kV, due to the small intensity, at least for a resolving power comparable to that of the contact microscope.

§ 7.   *The field of view.*

Another important magnitude of the microscope is the field of view. For, as the exposure times are large, as little as possible of the field of view should be wasted unnecessarily. If the maximum angle of the X-rays leaving the microscope is $2a$, the field of view, expressed as a length dimension, is $2a \tan a$ (see fig. 6). The number of image elements is proportional to the square of this field of view. By increasing the object distance n times, the field of view is also increased by a factor n. For the same magnification the film distance however has to be made n times larger, so that the exposure time required must be increased n^2 times, but so is the number of image elements obtained. Hence the "speed" of the

21

microscope, i.e. the number of image elements that can be studied per unit of time, at a given construction, is a constant, determined by the angle of view 2α (see fig. 6a and 6b).

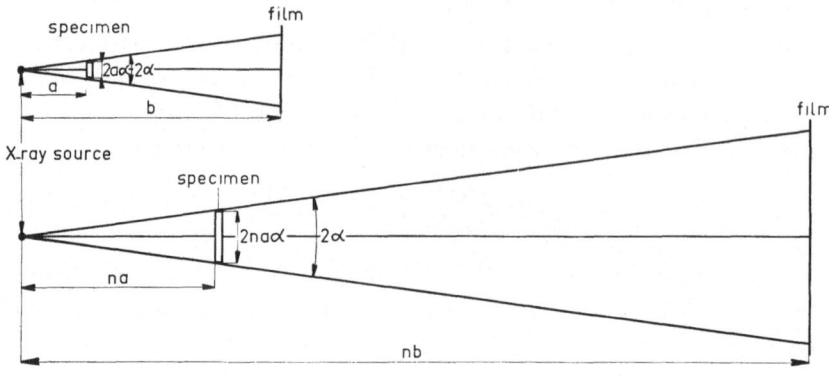

Fig. 6
Geometrical arrangement and field of view.

The angle of view is limited by two factors:

1) The angular intensity distribution.

At the film the intensity of the radiation decreases towards the edge. This is a consequence of 4 causes:

a) The shape of the angular emission distribution curve (see fig. 7). For a very thin target a minimum may occur in the axial direction and a maximum at an angle of 45°. This applies to white radiation. For characteristic radiation a spherical distribution may be expected, which can be explained from the mechanism for the excitation of radiation. White radiation originates by slowing down fast electrons. With a thin target, i.e. so thin that the electrons can penetrate it almost without scattering, the deceleration is parallel to the optical axis. Consequently the direction of the radiation is perpendicular to this axis. If the electrons are scattered several times before leaving the target, the electrons can also have a component of their velocity at an angle with the optical axis. With a very thick target the radiation characteristic is spherical again. The mechanism for the excitation of characteristic radiation is totally different. Here radiation is excited as a result of energy transitions in the atom. The direction of the corresponding electric field is completely

22

random. Measurements of radiation characteristics have been carried out bij Botden Combée and Houtman[4]) and further by Cosslett and Dyson[7]).

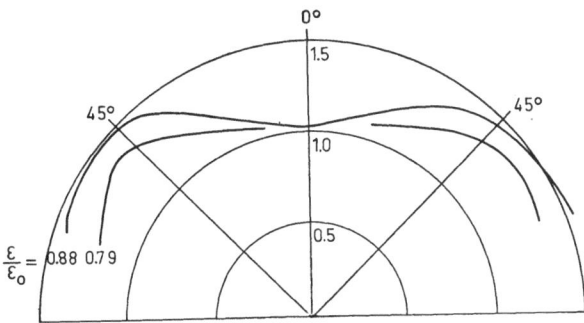

Fig. 7
Angular intensity distribution of the X-ray source according to Cosslett and Dyson[7]).

b) The effective absorbing thickness of the target increases towards the edge of the field. It follows from fig. 8 that

$$s_\alpha = \frac{s_o}{\cos \alpha} \tag{9}$$

in which s_α is the pathlength in the target for the X-rays that makes an angle α with the axis, and s_o is the thickness of the target.

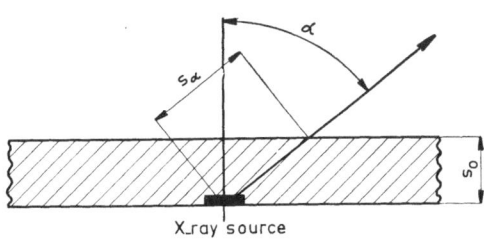

Fig. 8
The pathlength of the X-rays in the target.

c) The X-ray source to film distance increases towards the edge as a $\frac{1}{\cos \alpha}$ function (see fig. 9).

$$b_\alpha = \frac{b_o}{\cos \alpha} \tag{10}$$

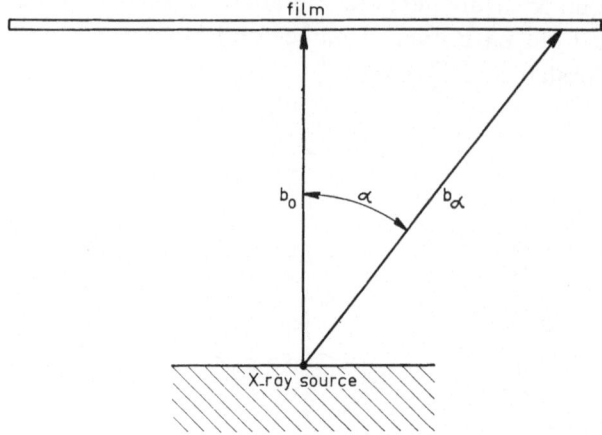

Fig. 9
The source to film distance.

d) The cosine effect. A radiation cone with a solid angle φ illuminates a smaller area in the middle of the film than at the edge; even if the distances have already been taken into account according to equation (10). According to fig. 10

$$q_\alpha = \frac{q_0}{\cos \alpha} \qquad (11)$$

in which q_0 is the area in the centre of the film and q_α the area at the edge that is illuminated by a radiation cone with solid angle φ, making an angle α with the optical axis.

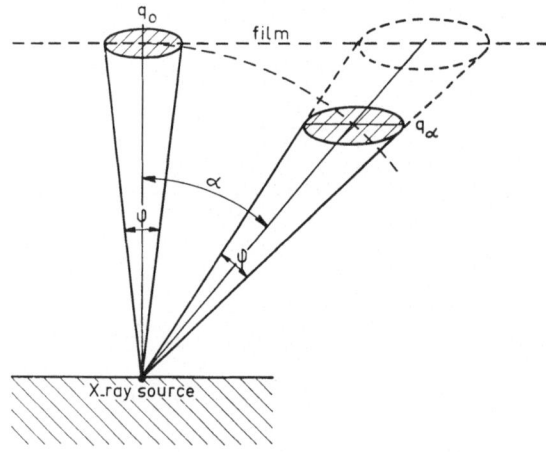

Fig. 10
The cosine effect of the illumination.

24

Thus the intensity continually decreases towards the edge. The total angle of view that can finally be used, cannot be expressed unambiguously in terms of a fixed value. Obviously with low contrast objects an even distribution is more important than with high contrast objects. In many cases the half angle of view amounts to $a \leqslant 15°$.

2) A material limitation.

In practice the upper pole piece of the lens forms a real limitation to the angle of view. In order to keep the spherical aberration small, a strong lens must be used. According to Liebmann and Grad[13]) for a strong lens

$$C_s \infty f^{5/3} \qquad (12)$$

holds, in which C_s is the spherical aberration constant and f the focal length of the lens. Further, a small focal length is also desirable in view of the chromatic aberration. (See also Van Dorsten and Le Poole[9])). A necessary consequence of a very strong lens is that the position of the focus is between the pole pieces (see fig. 11a), so that the maximum angle of view is limited by the bore of the upper pole piece.

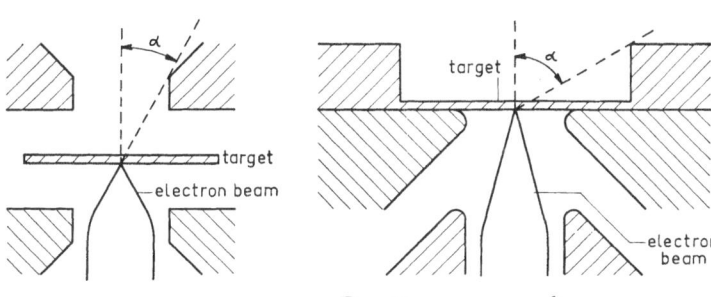

a   Fig. 11   b
Different constructions of the pole pieces showing the limitation of the angular field of view.

In a construction in which the upper pole piece is kept plane (fig. 11 b) the free angle of view can be very wide. To keep the focal length reasonably small the bore as well as the pole piece distance must be made proportionally small. However, because of the utilization of nearly the entire magnetic field, the value of C_s for a given focal length is larger than that for the construction represented in fig. 11a. Hence the construction of fig. 11b is less successful from an electron optical point of view. It has however, beside a wide angle of view, also the advantage that the object space is completely free. This construction was used mainly because the target is easily interchangeable during operation, which allows the possibility of studying an object successively with rays of different wavelengths (see Le Poole and Ong[15])), If the decrease in intensity towards the edge is not too large, some correction is possible in the photographic process.

§ 8. *Depth of penetration and diffusion of the electrons.*

According to § 6 the anode voltage must be kept low to obtain contrast. This is desirable also for other reasons. Von Ardenne[2]) mentioned the depth of penetration and the diffusion of the electrons in the target as a possible limitation of the minimum size of the X-ray source, and this was later confirmed experimentally by Nixon[14]). According to the Thomson-Whiddington[17]) equation the depth of penetration x_e of electrons in matter amounts to *)

$$x_e = \frac{AV^2}{CZ\varrho} \tag{13}$$

in which ϱ is the density, A the atomic weight, Z the nuclear charge, V the anode voltage, and C a constant. The spatial distribution of the electrons in the target is schematized in fig. 12.

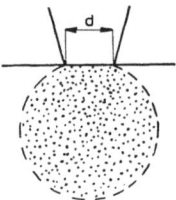

Fig. 12
Electron diffusion in the target.

Cosslett[8]) points out that probably X-rays are not excited in the entire field. If x_e, ϱ and V are expressed in respectively cm, g/cm^3 and Volt, C is about 6×10^{11}. For gold at 10 kV the depth of penetration is some 0.15 μ. By using a very thin target it is possible to

---

*) The Thomson-Whiddington law gives only a coarse approximation of the expected focus extention. At the Symposium on X-ray microscopy (Stockholm 1959), Hink showed that the effect of electron penetration is much smaller, at least for characteristic radiation. At the occasion Langner gave a more theoretical paper on electron diffusion.

reduce the influence of the effects of scattering. Nixon[14]) has succeeded in obtaining a resolution of 0.1 μ, using an 0.1 μ Au target
and 10 kV anode voltage.

§ 9. *Intensity of the X-ray source.*

The maximum brightness of an electron source, according to
Langmuir[12]), amounts to

$$\beta = \frac{j\,e\,V}{\pi\,k\,T} \qquad (14)$$

in which j is the saturation current density of the cathode, eV the
energy of the electrons, k Boltzmann's constant (1/11600 eV/°K)
and T the cathode temperature in °K. From the second law of ther-
modynamics it follows that the brightness of radiation in an optical system is constant, independent of the optical system used, or
$\frac{\beta}{n^2}$ is constant, in which β is the brightness and n the refractive index. If the lens aperture is γ the current density in the focus is

$$J = \pi\,\gamma^2\,\beta = \frac{\gamma^2\,j\,e\,V}{k\,T} \qquad (15)$$

At a focal diameter d the target current amounts to

$$I = \frac{\pi\,d^2\,\gamma^2\,j\,e\,V}{4\,k\,T} \qquad (16)$$

and the anode dissipation to

$$D = \frac{\pi\,d^2\,\gamma^2\,j\,e\,V^2}{4\,k\,T} \qquad (17)$$

The size of the aperture γ is limited by lens errors, of which only
chromatic and spherical aberration and astigmatism are important.
The chromatic error can be reduced sufficiently by stabilizing the
lens current and anode voltage. Although in principle correction
of astigmatism can be carried out easily if it can be observed visually, in practice it has not succeeded very well as yet. Spherical
aberration cannot or not easily be corrected, so it is this error
that determines the size of the aperture. The diameter of the disc
of confusion as a result of the spherical aberration in the paraxial
focus is given by

$$d_{sf} = 2\,C_s\,\gamma^3 \qquad (18)$$

27

in which C_s is the spherical aberration constant. Somewhere between lens and paraxial focus however, lies a plane with the smallest cross section. Here the diameter of the electron focus is

$$d_{\min} = {}^1\!/_2 C_s \gamma^3 \qquad (19)$$

Thus the lens aperture is determined by the following equation

$${}^1\!/_2 C_s \gamma^3 \leq d \qquad (20)$$

Equalizing both sides of (20) gives, together with (17), for the anode dissipation

$$D = \frac{\pi \, d^{8/3} \, j \, e \, V^2}{4 \, k \, T \, ({}^1\!/_2 \, C_s)^{2/3}} \qquad (21)$$

For the excitation of X-rays the energy efficiency for white radiation is given by

$$\eta_x = 10^{-9} \, V \, Z \qquad (22)$$

in which Z is the nuclear charge. Thus the total X-ray energy amounts to

$$E_x = C \, V^3 \, d^{8/3} \, C_s^{-2/3} \qquad (23)$$

in which

$$C = \frac{10^{-9} \, Z \, \pi \, j \, e}{2.5 \, k \, T} . \qquad (24)$$

§ 10. *Heat generation in the target.*

Almost all the energy according to equation (21) is transformed into heat. With normal X-ray tubes with a focal diameter of the order of magnitude of mm the heat flow forms the largest problem. As the focal size decreases the specific anode load may be increased without the local temperature becoming inadmissibly high[5][16]. With a small round focus, as is used in the projection X-ray microscope, we can consider the heat flow to be purely radial (see fig. 13).

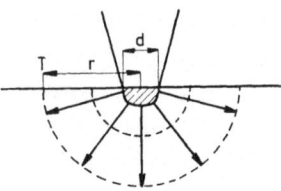

Fig. 13
Heat flow in the target.

For a certain specific loading W the total anode dissipation is

$$D = \frac{\pi}{4} W d^2 \qquad (25)$$

We can calculate the temperature T_t at a distance r of the focal centre from

$$m \frac{d T_t}{d r} 2 \pi r^2 = - \frac{\pi}{4} W d^2$$

or

$$T_t = \frac{W d^2}{8 m r} + T_o \qquad (26)$$

which however only holds for $r \geqslant \frac{d}{2}$.

In equation (26) m is the thermal conductivity. For $r = \frac{d}{2}$ the temperature rise amounts to

$$T_d = \frac{W d}{4 m} \qquad (27)$$

When using a certain voltage with the projection X-ray microscope (cf. equation 21)

$$W d^2 = \frac{\pi d^{8/3} j e V^2}{4 k T (1/2 C_s)^{2/3}} \qquad (28)$$

This gives with (27)

$$T_d = \frac{\pi d^{5/3} j e V^2}{16 k T m (1/2 C_s)^{2/3}}. \qquad (29)$$

Assuming an admissible temperature rise $T_d = 200°$ C and inserting $V = 10^4$ V, $T = 2600°$ K, $k = 1/11600$ eV/$°$ K, $j = 5.10^{-3}$ A/mm^2, $C_s = 0.5$ mm., $m = 0.20$ W/mm.$°$ C, we get $d \approx 2\ \mu$. For $d < 2\ u$ the maximum target temperature becomes even lower according to (29). Consequently heat generation does not form a limiting factor in the projection X-ray microscope yet.

§ 11.   *Practical limitation of the resolving power.*

The quality of both optical and electron microscopes is characterized mainly by resolving power and size of the field of view. The brightness of the image does play an important part in obtaining resolution, especially in the high resolution electron microscope, but exposure times still lie within reasonable limits. Visual focusing is still possible at 10 Å resolution.

The quality of the X-ray projection microscope however, depends strongly on a third factor: the wavelength of the radiation used. When using white radiation the wavelength is determined by the anode voltage, so that the quality of the X-ray images improves with decreasing anode voltage. The X-ray intensity however decreases according to equation (23) with the cube of the voltage. Assuming complete absorption of the radiation in film or screen, the energy according to (23), determines the exposure time and also the possibility of focusing visually. At a given voltage and resolution the total X-ray output can be increased only by making j/T larger or C_s smaller. The factor $V^3 d^{8/3}$ indicates how to utilize this gain in energy, i.e. either for decreasing the anode voltage or for improving the resolving power when using the same exposure time. The question rises whether, at given values of C_s and j/T, the quality of the microscope can be improved by decreasing the voltage and/or the source diameter at the expense of a longer exposure time. This is indeed possible as long as a minimum X-ray output is not reached, which in practice is determined rather by the impossibility of visual focusing than by the excessive exposure time. This can be elucidated as follows (c.f. chapter IV): The visibility of a detail is determined by the number of X-ray quanta used for building up an image element. At a certain screen distance this number is proportional to the X-ray intensity and the storage time of the eye or the fluorescent screen. Though when using a film, exposure time is not limited, in practice an exposure time of some 20 mins. must be considered as the maximum in view of the stability of the apparatus. Supposing the storage time of the eye to be 0.1 sec, this means an improvement of the signal to noise ratio with respect to the direct observation of $\sqrt{12000}$ or roughly 100 times. It is true that during focusing a gain in visual brightness can be obtained at the expense of the field by bringing the test object very close to the target, as in fact one image element would be sufficient for focusing, but manipulations on the vulnerable target are most undesirable. Consequently we must conclude from these considerations that in practice the minimum X-ray intensity is determined by the impossibility of visual focusing.

§ 12. *Optical aids for focusing.*

As the quality of the image depends so much on the accuracy with which the apparatus is focused, it is desirable to pay due attention to this problem. By using an optical aid a considerable gain in brightness can be obtained. For focusing a specially suited object is used, usually a very fine grid of some heavy element. The image on the screen is viewed and the lens current varied until the sharpest image is obtained. As soon as the apparatus is focused, the test object is replaced by the object to be studied, while keeping the lens current and anode voltage constant.

When focusing with the unaided eye the magnification on the screen must satisfy

$$M \geq \frac{\delta_{eye}}{\delta} \qquad (30)$$

in which δ_{eye} and δ are the resolution of the unaided eye and the microscope respectively. If we use an optical system with a magnification M' for observing the fluorescent image a magnification on the screen of

$$M \geq \frac{\delta_{eye}}{M' \delta} \qquad (31)$$

is enough. The magnification on the screen M is proportional to the screen distance b while the brightness of the fluorescent image is inversely proportional to the square of this distance. This means that when using an optical system with magnification M' the image is M'^2 times brighter, provided the pupil of the eye is completely filled. Since the useful magnification of the optical system is limited by the resolution of the screen the gain in brightness depends on its grain size. The maximum useful magnification is about 20 times. This corresponds to a 400 times gain in brightness. This argument only holds if the screen is viewed with a binocular with two separate objectives (stereomicroscope).

§ 13. *Conclusions and directives for the development of the projection X-ray microscope.*

The conclusions we can draw from the considerations in this chapter are as follows:
1) The quality of the projection X-ray microscope image is mainly limited by the X-ray output.
2) The limitation of the X-ray output is not the result of inadmissible heat generation in the target, but of the limited brightness of the electron source, combined with image errors of the electron lens and the desirability of the use of low voltages.

3) The impossibility of visual focusing forms a limitation rather than excessive exposure time.

Conclusions 2 and 3 also give directives for developing the microscope. Conclusion 2 gives the causes of the limited X-ray output. If the electron source is properly designed, improvement according to equation (14) is only possible when using better cathode materials with a more favourable value of j/T.

It is questionable, however, whether in our experiments we obtain the maximum theoretical brightness. Although the author has done some theoretical work on this matter, no experiments have been carried out up to now.

Furthermore it is useful to reduce lens errors as far as possible. To this end currents and voltages can be stabilized, and astigmatism corrected, at least in principle. Spherical aberration however is very difficult to correct, it can only be reduced by a proper design of the electron lens. During development no effort was made towards actual lens correction.

Besides low anode voltages being desirable to get good contrast, they are necessary to keep the depth of penetration of the electrons into the target small. The effect of penetration on resolution can be reduced by using a very thin target. A suitable spectral distribution can still be obtained when utilizing characteristic radiations and relatively high voltages.

Contrast can be further improved by staining or shadowing the object. These possibilities are studied in chapter VI. Furthermore contrast can be improved considerably by correct choice and use of the film. This is further studied in chapter IV and V. Conclusion 3 expresses the desirability of using some focusing aid which does not suffer so much from the low X-ray intensity. The focusing method developed in this laboratory and described in chapter III proved satisfactory in practice.

REFERENCES

1) Ardenne, M. v.   Tabellen der Elektronenphysik, Band I, Deutscher Verlag der Wissenschaften, Berlin (1956) p. 439.

2) Ardenne, M. v.   Elektronen Uebermikroskopie, Verlag von Julius Springer, Berlin (1940) p. 73.

3) Baez, A. V. and M. A. El-Sum   X-ray microscopy and microradiography, Academic Press inc. New York (1957) p. 347.

4) Botden, P. J. M., B. Combée and J. Houtman   Phil. Tech. Rev. **14** (1952) 114.

5) Cosslet, V. E. Proc. Phys. Soc. B **LXV** (1952) 782.

6) Cosslett, V. E. and Nature, **168** (1951) 24.
W. C. Nixon

7) Cosslett, V. E. and X-ray microscopy and microradiography, Academic
N. A. Dyson Press inc. New York (1957) p. 405.

8) Cosslett, V. E. Proc. Third. Int. Conf. on Elec. Micr. London (1954)
p. 311.

9) Dorsten, A. C. van Phil. Tech. Rev. **17** (1955) 47.
and J. B. Le Poole

10) Engström, A., X-ray microscopy and microradiography, Academic
R. C. Greulich, Press inc. New York (1957) p. 218.
B. L. Henke and
B. Lundberg

11) Gabor, D. Proc. Roy. Soc. **197** (1949) 454.

12) Langmuir, D. B. Proc. I.R.E. **25** (1937) 977.

13) Liebmann, G. and Proc. Phys. Soc. B **64** (1951) 56.
E. M. Grad

14) Nixon, W. C. Nature, **175** (1955) 1078.
Proc. Roy. Soc. A **232** (1955) 475.

15) Ong Sing Poen and X-ray microscopy and microradiography, Academic
J. B. Le Poole Press inc. New York (1957) 91.

16) Oosterkamp, W. J. Phil, Res. Rep. **3** (1948) 49.,
Ibid, **3** (1948) 161., Ibid, **3** (1948) 303.

17) Whiddington, R. Proc. Roy. Soc. A **89** (1914) 554.

CHAPTER III

A NEW FOCUSING AID

§ 1. *Introduction.*

As the resolution of the projection X-ray microscope is limited by the focusing accuracy (c.f. chapter II), a more critical and effective focusing method was sought in the course of the work. The difficulty of the conventional method is mainly due to the low intensity combined with a low contrast of the image. Consequently the object under examination is generally unsuitable for focusing purposes. For this reason a more suitable fine mesh grid is commonly used. With the low intensities available, the screen distance has to be kept as small as possible, which in turn requires the grid to be in close contact with the target. The screen brightness will then be determined by the smallest grid-source distance and the resolution, as for higher resolution a larger magnification is imperative (c.f. chapter II). In practice it is hardly possible to make a reproduceable minimum grid to source distance smaller than $a \approx 30 \ \mu$, mainly due to the unevenness of the grid and the curvature of the target under atmospheric pressure (see fig. 1a). Of course it is possible to fix the grid on the target (fig. 1b), thus minimizing the distance a. But in this case the microscope has always to be used together with the fixed testgrid.

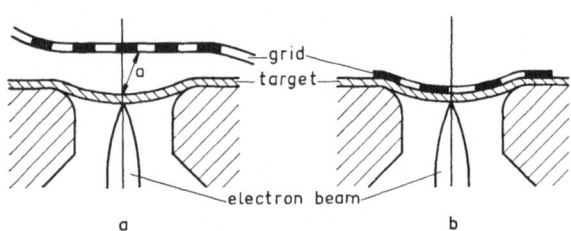

Fig.1 a and b

Position of the test grid with respect to the target. In case b the grid is fixed on the target.

Accurate adjustment of the source with respect to the grid position is necessary to avoid the entire image falling within the shadow of a mesh.

The use of image amplification can give some intensity improvement. However, although it can bring the brightness to a conveniently high level, so that dark-adaption of the eye is unnecessary, it will not improve the contrast, since with the same storage time, instead of improving the signal to noise ratio, it will always introduce extra noise. So the problems concerning the focusing accuracy remain. Here follow some of the important problems:

1) To shorten the distance a, the testgrid has to be thin, resulting in a low contrast X-ray image and hence inaccurate focusing.

2) Focusing and exposure are two separate operations, so that changes between these two operations cannot be detected; neither can this be done during exposure.

3) The focusing procedure is inconvenient and time-consuming, so that the microscope is not suitable for quick routine work.

4) Due to the low intensity, mainly resulting in low contrast if an amplifier is used, this method cannot be used for high resolution and/or low voltages.

§ 2. The new focusing aid.

The focusing aid which will be described here has more pleasant properties and reduces the difficulties to a large extent.

It makes use of the fact that some of the electrons striking the target are elastically reflected. The energy distribution of the total secondary emission for a primary energy of about 155 V is given in the literature[3] [5] [10] [11]; see fig. 2. It has a sharp peak for an

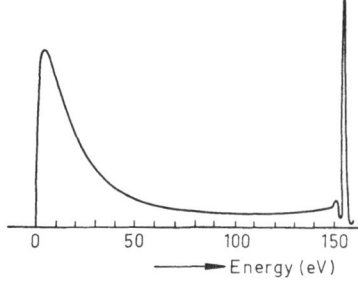

Fig. 2. Energy distribution of secondary electrons for a primary energy of 155 eV according to Rudberg[10] [11]

energy equal to that of the primary electrons. The secondary electrons enter the lens from the opposite direction. Some of them pass the aperture, and those which are elastically reflected form a magnified and sharp secondary image at the electron source.

If the instrument is perfectly aligned, this image will be exactly at the electron source. By introducing a transverse magnetic field (fig. 3a) the returning beam can be separated from the primary beam, thus allowing observation of the secondary image on a fluorescent screen. Le Poole proposes to obtain this transverse field by slightly tilting the objective lens, if this is of the magnetic type. In this case there is a field component perpendicular to the optical axis.

If the lens has no errors and the target is in focus, the secondary image will be equal in size and shape to the electron source. As its diameter is about 40u, an optical magnification of some 10 times is required to obtain the necessary information from this image.

§ 3. *Separation of the beams.*

The separation of the primary and the returning beam by tilting the lens can be considered as caused by the rotation of the lens and will now be examined more closely. In fig. 4 the magnetic lens

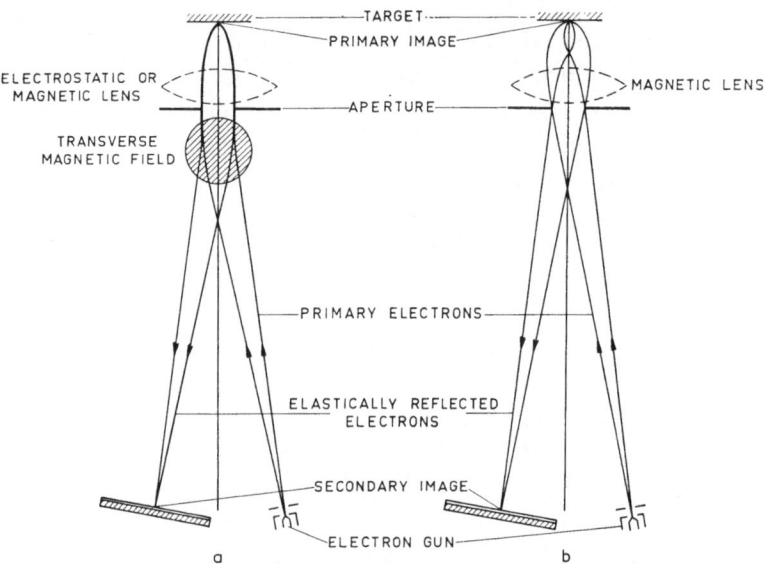

Fig. 3. Separation of the primary and returning beams.

is represented by a cylinder. The direction of the magnetic field is indicated by the arrow B. The image rotations for electrons coming from C and T are in the directions indicated by D_C and D_T respectively. As can be derived from fig. 4, D_C and D_T have the same sense of rotation.

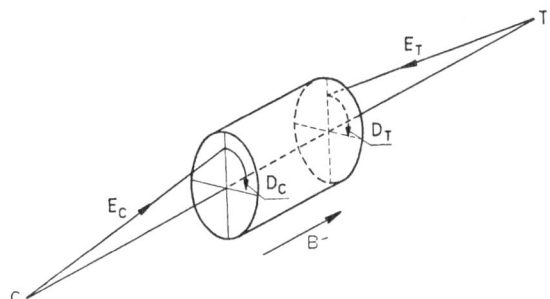

Fig. 4. Elucidation of the beam separation caused by image rotation.

To elucidate the separation of the rays caused by this rotation we examine the projection on a plane perpendicular to the optical axis (fig. 5b). The lens is now represented by a circle and the optical axis by the centre A. Instead of tilting the lens the electrons are directed obliquely to the lens. On doing so, the electron source

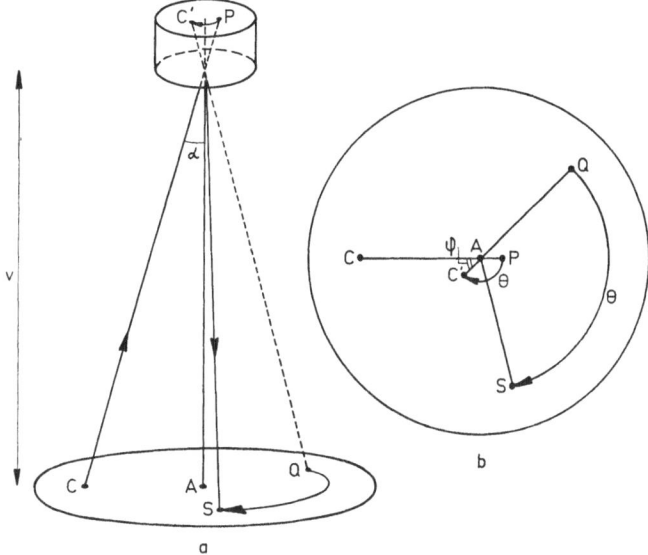

Fig. 5. Construction of the projection diagram (fig. 5b).

is represented by C, not coinciding with the optical axis. If there were no rotation, the primary image on the target would be represented by P, CA/AP being the demagnification. Due to the rotation by the angle θ the projection of the primary image comes in C'. Without rotation the elastically reflected electrons coming from C' would give a secondary image in Q. With rotation in the same direction and by the same angle θ the secondary image now comes in S, SA/AC' = CA/AP being the magnification.

In our projection diagram the angle CAS = 2φ = 2 (180° — θ). Further we can derive that

$$CS = 2\,CA\,\sin \varphi = 2\,CA\,\sin (180° — \theta). \tag{1}$$

CS will be zero, that is, the secondary image S coincides with the electron source if CA = 0 or if $\theta = n \times 180°$, $n = 0, 1, 2, \ldots\ldots\ldots$

CA = 0 means that the electron source is situated on the optical axis and consequently that the lens is exactly aligned. Since it can be proved that $0 < \theta < 180°$, the beams are always separated when the lens is not exactly aligned. If in this case the lens power is varied, the secondary image S besides getting blurred will describe an arc. If S coincides with C, that is, if the lens is centered, the displacement according to (1) is zero and the image is only blurred.

§ 4. *Accuracy of alignment.*

In order to avoid inadmissible image errors the angle of tilt a between the optical axis of the lens and that of the electron source must be kept small. From fig. 5a it can be derived that

$$a = CA/v = CS/2v \sin \varphi. \tag{2}$$

At a given rotation angle θ and distance v between source and lens the distance CS between electron source and secondary image must therefore be small. To realize this the focusing screen must be placed in the primary beam and provided with a hole to allow the primary electrons to pass. The smallest distance CS is then determined by the radius of this hole. At a given value of CS the tilting angle is a minimum if $\sin \varphi = 1$ or $\varphi = 90°$ and thus $\theta = 90°$. Furthermore (2) shows that the alignment improves with increasing v. Note that the projection diagram as shown in fig. 5b can in fact represent the focusing screen if we assume C to be fixed at the hole in the screen. In this case the centre A is displaced by tilting the lens.

In our microscope $CS = 1$ mm, $v = 400$ mm, $\varphi = \approx 60°$ [2]), $\sin \varphi = \frac{1}{2} \sqrt{3}$. So $\alpha < 1/600$ radians. Note that this angle is much smaller than can be obtained by centering by means of the X-ray image in the conventional way. In our apparatus this critical centering characteristic is used also for alignment of the condenser lens. See also § 10.

§ 5. *Focusing accuracy.*

As in this device the errors are repeated on the return, the focusing accuracy can be high. Let us investigate what happens if the target is not in focus. In fig. 6 C represents the electron source, T the target, B the paraxial focal plane and L the objective. If the focal length of the lens is f, we may write (see fig. 6) according to the paraxial lens equation

$$v = fb/(b - f). \tag{3}$$

The radius of the blur on the target equals

$$\varepsilon = (t - b) \frac{h}{b} = \left(\frac{t}{b} - 1 \right) h. \tag{4}$$

The secondary image of the blur is sharp in M at a distance v' from the lens, in which

$$v' = ft/(t - f). \tag{5}$$

Apart from the effect of changes in the intensity distribution, the blur at the electron source will have a radius

$$\varrho = 2(v - v') \frac{h}{v'} = 2h \left(\frac{v}{v'} - 1 \right).$$

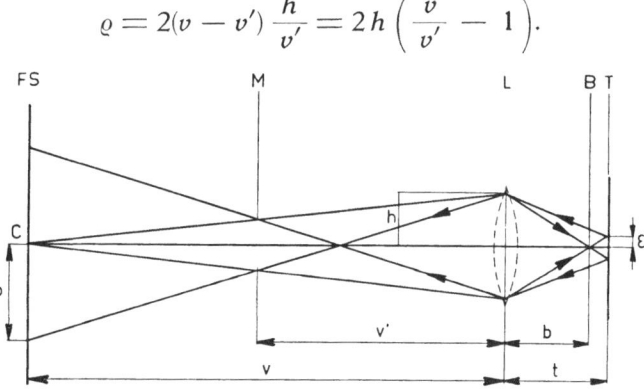

Fig. 6. The unsharpness of the secondary image due to incorrect fosusing. FS = focusing screen, C = electron source, M = image plane of the secondary image, L = objective lens, B = image plane of the primary image, T = target, ϱ = radius of the unsharpness on the focusing screen, ε = radius of the unsharpness of the primary image.

Substitution of (5) gives

$$\varrho = 2h\left(\frac{v}{f} - \frac{v}{t} - 1\right). \tag{6}$$

Putting the expression in parentheses equal to zero gives the well-known lens equation

$$\frac{1}{v} + \frac{1}{t} = \frac{1}{f}. \tag{7}$$

Assuming a Gaussian intensity distribution, a more accurate calculation shows that instead of (6) we should write

$$\varrho = \sqrt{2}\,h\left(\frac{v}{f} - \frac{v}{t} - 1\right). \tag{8}$$

In the neighbourhood of the focusing point $\varrho = 0$ we may write

$$\varrho = \varrho_0 + \left(\frac{\partial\varrho}{\partial f}\right)_0 \Delta f + \left(\frac{\partial^2\varrho}{\partial f^2}\right)_0 (\Delta f)^2 + \ldots. \tag{9}$$

For small values of Δf we may neglect the terms of second and higher power in Δf and taking into account that $\varrho_0 = 0$ we get

$$\varrho = \left(\frac{\partial\varrho}{\partial f}\right)_0 \Delta f. \tag{10}$$

According to (8)

$$\frac{\partial\varrho}{\partial f} = -\frac{\sqrt{2}\,hv}{f^2}. \tag{11}$$

With (7) this becomes

$$\left(\frac{\partial\varrho}{\partial f}\right)_0 = -\frac{\sqrt{2}\,h\,(t^2 + 2tv + v)}{vt^2}.$$

In general $v \gg t$; so we may write

$$\left(\frac{\partial\varrho}{\partial f}\right)_0 = \frac{-\sqrt{2}\,hv}{t^2} = \frac{-\sqrt{2}\,hM}{t}, \tag{12}$$

in which
$$M = v/t \tag{13}$$

is the demagnification for the primary image and magnification for the secondary image. Further $M \gg 1$ and $t \approx f$, which gives for (12)

$$\left(\frac{\partial\varrho}{\partial f}\right)_0 = -\sqrt{2}\,hM/f \tag{14}$$

and for (10)
$$\varrho = -\sqrt{2}\,hM\,\Delta f/f \tag{15}$$

40

Putting the resolution δ equal to the diameter of the primary image d we get according to (13)

$$M = c/d = c/\delta, \tag{16}$$

in which c is the diameter of the electron source. Equation (15) can then be transformed to

$$\frac{\varrho}{c} = - \sqrt{2} \; \frac{h}{\delta} \frac{\varDelta f}{f}. \tag{17}$$

The radius of the blur in the primary image amounts to (fig. 6)

$$\varepsilon = h \varDelta f/f. \tag{18}$$

Inserting this in (17) gives

$$\varrho/c = - \sqrt{2} \; \varepsilon/\delta, \tag{19}$$

in which ϱ/c is the relative error in the secondary image and ε/δ the relative error in the primary image. For focusing we have to satisfy the condition $|\varepsilon/\delta| < 1$ and so the condition for good focusing becomes

$$|\varrho/c| < \sqrt{2}. \tag{20}$$

As shown, this is independent of the resolution δ. The minimum value of $|\varrho/c|$ is determined by the lens errors only.

§ 6. *Position of the focusing screen.*

It was shown that the secondary image is in the plane of the source. As this place is inaccessible, we shall investigate the admissible distance $\varDelta v$ between source and screen. In fig. 7 FS represents the focusing screen. With the target in focus the radius of the blurring at FS equals

$$\varrho_s = \frac{\varDelta v}{v} \, h. \tag{21}$$

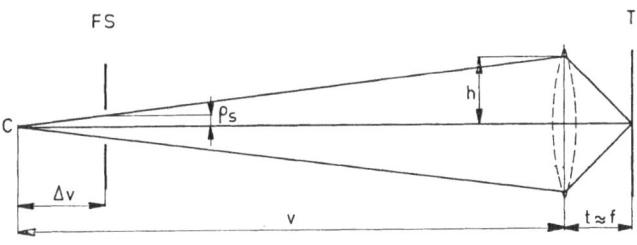

Fig. 7. Position of the focusing screen. Due to the great depth of focus it is not necessary to place the focusing screen at the electron source level.
T = target, C = electron source.

The half-angle aperture of the objective is $\gamma = h/f$. Further $v/f \approx M$ and $M = c/\delta$. So

$$\varrho_s = \frac{\varDelta v}{c} \, \delta \gamma. \tag{22}$$

We have to ensure that $\varrho_s/c < 1$ and thus

$$\varDelta v < c^2/\delta \gamma \tag{23}$$

In our microscope we use $\gamma \approx 0.08$ for $\delta = 0.1\,\mu$ and $c \approx 40\,\mu$. So the condition (23) becomes $\varDelta V < 200$ mm. With increasing δ a more correct position of the screen is required.

§ 7. *Brightness of the secondary image.*

In practice the secondary image has proved to be sufficiently bright, even when the X-ray intensity is so low that focusing by the conventional method is impossible. This however is only partly due to the fact that the reflection coefficient for the electrons is greater than the X-ray efficiency. The main reason is that the electrons are focused to a small spot while the X-rays diverge to illuminate a large plane. In other words the method corresponds to the use of only one image element as mentioned in chapter **II** § 11.

In fig. 8 **T** represents the target, **RS** the X-ray fluorescent

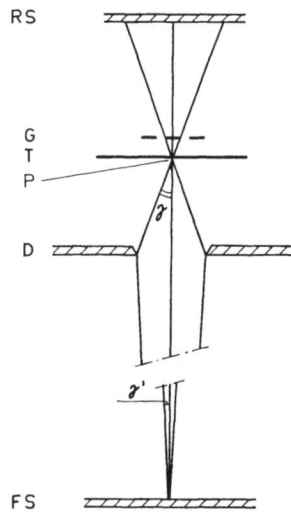

Fig. 8. Comparison of the intensities on the X-ray screen and focusing screen.
RS = X-ray screen, G = test grid, T = target, P = primary image,
D = diaphragm, FS = focusing screen.

screen, G the test grid, P the primary electron image and FS the focusing screen.

The current density in P amounts to

$$j_p = \pi \gamma^2 B, \tag{24}$$

in which B is the brightness of the electron source. If the reflected electrons follow Lambert's law, the brightness for the secondary radiation is

$$B_s = \eta_e \gamma^2 B, \tag{25}$$

in which η_e is the reflection coefficient. The current density of the secondary image on the focusing screen FS therefore amounts to

$$j_s = \pi \eta_e \gamma^2 \gamma'^2 B, \tag{26}$$

and with $\gamma' = \gamma/M$ and $M = \delta/c$ this becomes

$$j_s = \pi \eta_e \gamma^4 (\delta/c)^2 B. \tag{27}$$

The half-angle γ is determined by spherical aberration to be

$$\gamma = (2\delta/C_s)^{1/3}, \tag{28}$$

C_s being the spherical aberration constant.

Equation (27) then becomes

$$j_s = \frac{\pi \eta_e \delta^{10/3} B}{(^1/_2 \, C_s)^{4/3} c^2} \tag{29}$$

Hence the current density is inversely proportional to the 4/3 rd power of the spherical aberration constant C_s. As far as we know this is the only example in electron optics where the spherical aberration constant plays such an important rôle. Assuming a current density in the primary image of $j_p = \pi \gamma^2 B = 0.5 \text{ A/mm}^2$ we get by inserting into (27): $\eta_e = 2.3 \times 10^{-3}$ *), $\gamma = 0.08$, $\delta = 0.1 \, \mu$, $c = 40 \, \mu$, $j_s = 4 \times 10^{-11} \text{ A/mm}^2$. Assuming an afficiency of the fluorescent screen of $\eta_l = 30 \text{ lm/w}$, the brightness of the visual secondary image becomes $B_l = 7.2 \text{ nt}$. We shall now compare the intensity of the secondary image and the X-ray fluorescent image. By multiplying equation (27) by the anode voltage V we get

$$I_e = \pi \eta_e \, \gamma^4 B \left(\frac{\delta}{c} \right)^2 V \tag{30}$$

---

*) The value of η_e depends on the electron energy and target material. According to Rudberg[10][11] we get for gold at 150 V, $\eta_e = 2.5$ to 5%[2]). We take here the value of η_e a factor 10 lower for 6 kV.

in which I_e is the intensity of the secondary image. According to equation (24) and assuming an X-ray efficiency of η_x, the total X-ray energy amounts to

$$E_x = \frac{\pi^2}{4} \gamma^2 B \, \eta_x \, \delta^2 \, V \tag{31}$$

The X-ray intensity at the screen will be

$$I_x = \frac{\pi}{16} \gamma^2 B \eta_x V \left(\frac{\delta}{b}\right)^2 \tag{32}$$

b being the screen to source distance.
Consequently

$$\frac{I_e}{I_x} = 16 \frac{\eta_e}{\eta_x} \left(\frac{b}{c}\right)^2 \gamma^2 \tag{33}$$

This equation can be expressed in the more fundamental magnitudes δ, a and C_s. For the conventional focusing method we need, according to equation (30), chapter II, a magnification $M \approx \dfrac{b}{a} \approx \dfrac{\delta_s}{\delta}$ in which δ_s is the resolution of the screen. For the new method we can, if necessary, by adding an extra lens adapt the diameter of the secondary image to the screen resolution δ_s, so

$$\delta_s \approx c \tag{34}$$

For equation (33) we now may write

$$\frac{I_e}{I_x} = 16 \frac{\eta_e}{\eta_x} \frac{a^2}{d^2} \gamma^2 \tag{35}$$

and with $\gamma = \left(\dfrac{2\delta}{C_s}\right)^{1/3}$ (equation (28), this chapter) we now have

$$\frac{I_e}{I_x} = 25 \frac{\eta_e}{\eta_x} \frac{a^2}{\delta^{4/3} C_s^{2/3}} \tag{36}$$

Inserting $C_s = 500 \, \mu$, $a = 30 \, \mu$, $\delta = 0.1 \, \mu$ we get $I_e/I_x \approx 10^4$.
As can be seen from equation (36) the ratio I_e/I_x becomes better for high resolution.

It is true that the intensity of a plane and that of a spot are not strictly comparable and that we have neglected the effect of the background. On the other hand the X-ray image lacks contrast too, especially when using the thin test grid, while the reflection method yields a more critical indication, because the electrons pass the optical system twice.

§ 8. *Influence of inelastically scattered electrons.*

A relatively large number of the secondary electrons lose more or less energy, which results in a more or less bright background. Due to the sharp peak of the elastically reflected electrons, the actual secondary image is clearly distinguishable. The electrons with small energy loss even give rise to an effect that we can use for preliminary centering purposes.

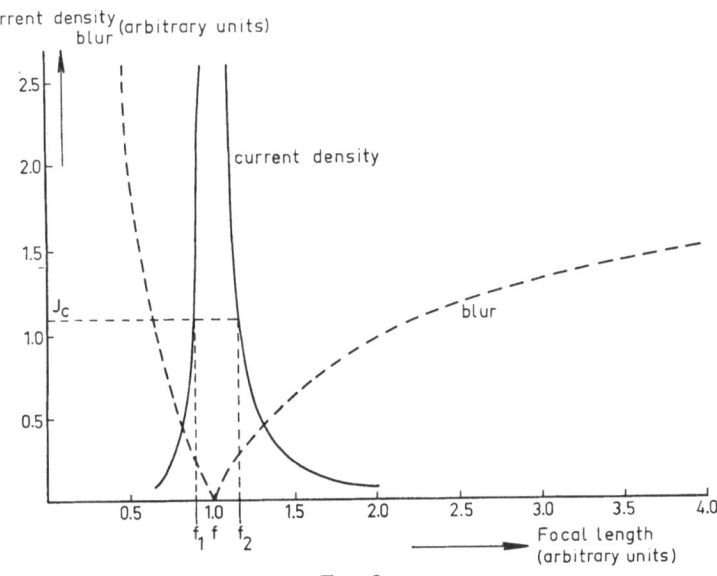

Fig. 9

Current density j and blur ϱ of the secondary image plotted against the focal length of the objective lens.

In fig. 9 the diameter of the secondary image (for an ideal lens) is plotted against the focal length according to equation (6). In the same diagram the corresponding relative current density is drawn. Assuming that a certain intensity level, corresponding to a current density j_c can just be detected on the screen, for an image to be seen the focal length has to be nearly correct. Due to the presence of electrons scattered with small energy losses there is a visible image even if the microscope is much further underfocused.

We want to go further into this matter, and therefore consider the 3 conditions: a) underfocused, b) focused, c) overfocused, in which the lenses have the respective focal lengths $(f + \Delta f)\uparrow$, $f\uparrow$ and $(f - \Delta f)\uparrow$. The upward arrows denote values applying to the electrons, that are directed upwards in fig. 10, that is, the primary electrons.

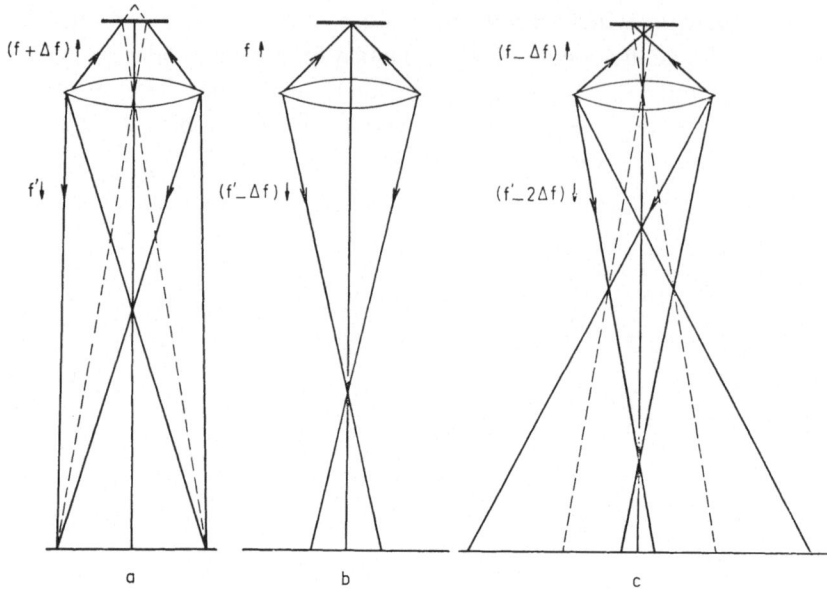

Fig. 10
The effect of the inelastically scattered electrons on the secondary image.

First we consider only that part of the returning radiation with an energy loss ΔeV and such that for this radiation the return image in case a) is good again. Suppose the focal length is $f' \downarrow$, in which the downward directed arrow indicates that this holds for return radiation. In cases b) and c) the focal lengths are respectively: $(f' - \Delta f) \downarrow$ and $(f' - 2\Delta f) \downarrow$. Those electrons with energy loss ΔeV give in case a) a sharp image of the blur of the primary image on the focusing screen. In case b) the blur is caused by the error Δf and in c) by an error twice as large. It is clear that in case c) the inelastically scattered electrons hardly contribute to the formation of the image as there are no electrons with an energy larger than eV. In case b) contribution can be given by those electrons whose energy losses are small, but their number is relatively very small. In case a) almost all electrons with an energy between eV and $(eV - \Delta eV)$ contribute to an image although a blurred one. The foregoing would suggest that the transition of position a) into b) is not sharp. This however has no detrimental influence on our focusing characteristic.

§ 9. *Magnetic stray field.*

In § 4 it was shown that correct alignment can be obtained by having the secondary image coincide approximately with the electron source. This, however, is only true if there is no transverse magnetic field between electron source and objective lens.

We shall now calculate the maximum field intensity which gives a negligible effect. Assuming there is a weak magnetic field B perpendicular to the plane of the diagram (fig. 11) and acting from the source C over a distance l we may write (see fig. 11) CS $= \xi l$ and $\xi = l/R$ with $R = 3.38 \times 10^{-6} V^{1/2}/B$, where CS, l and R are in m, V in V, B in Wb/m². So CS $= l^2 B/3.38 \times 10^{-6} V^{1/2}$. We require CS $< 10^{-3}$ m and get the condition

$$B < 3{,}38 \times 10^{-9} V^{1/2}/l^2 \qquad (37)$$

In our case $l = 0.4$ m, $V = 6.10^3$ V, we have to fulfill $B < 1.6 \times 10^{-6}$ Wb/m².

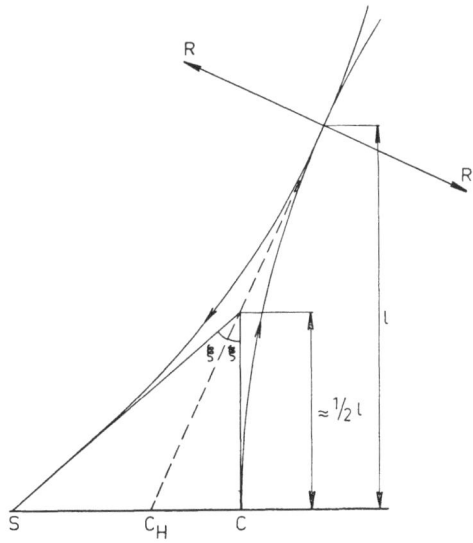

Fig. 11. Computation of the admissible transverse magnetic stray-field. C = electron source, C_H = the apparent electron source, S = secondary image, R = radius of the electron beam curvature.

In the presence of a stray field it is still possible by correctly tilting the objective to have the secondary image coincide with the source. This means that no conclusion as regards the alignment can be drawn from this coincidence.

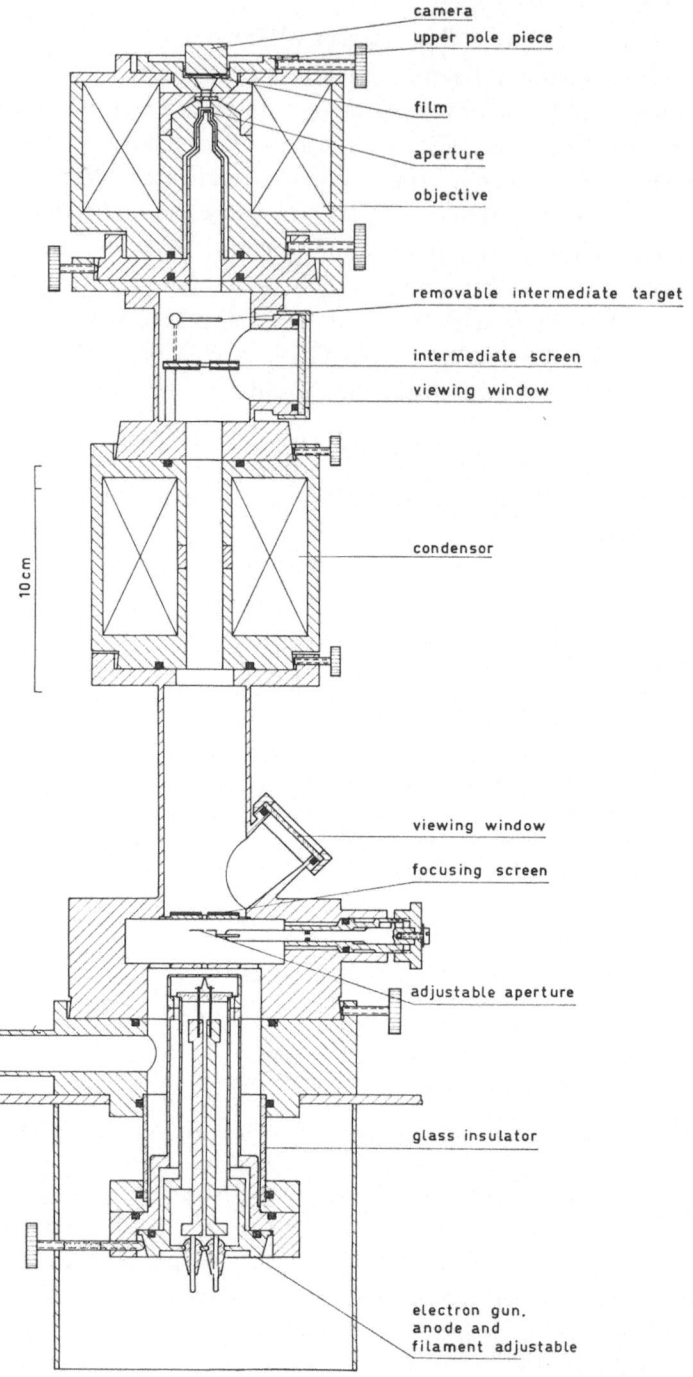

camera

upper pole piece

film

aperture

objective

removable intermediate target

intermediate screen

viewing window

10 cm

condensor

viewing window

focusing screen

adjustable aperture

glass insulator

electron gun,
anode and
filament adjustable

Fig. 12. Simplified cross-section of the experimental X-ray microscope.

Fig. 13. The Delft X-ray projection microscope, constructed early in 1957. Binocular observation of the focusing screen. In its present form the tilting is omitted and a 30 × monocular microscope is used for focusing (cf. fig. 12).

§ 10. *Practical performance.*

An improvised arrangement proved the intensity of the secondary image to be amply sufficient for focusing. One of our first photographs, using this focusing aid, was taken with such a low X-ray intensity that for correct exposure it took 1 to 2 hours at 4 kV and 6 mm film distance.

From this provisional apparatus valuable information has been obtained for an experimental microscope in which this focusing aid is incorporated. Fig. 12 shows a simplified cross-section of this microscope.

Fig. 14.   Close-up of the upper pole piece.

Immediately under the focusing screen there is an aperture allowing only a narrow beam to enter the microscope. This is necessary to suppress the background caused by reflection at the diaphragm.

An intermediate screen and a removable intermediate target are provided between objective and condenser. The screen enables us to catch the secondary image even when the lens is tilted too much. The intermediate target is used to centre the condenser. Electrical insulation makes it possible to measure the electron current if necessary. So these two parts are only used for preliminary alignment. The objective aperture can be centered during operation as a high accuracy is required. Previously the objective could also be tilted, but this proved impractical since tilting and shifting in fact have the same effect. In our present apparatus the tilting is therefore omitted, as shown in fig. 13. It was found that positioning of the pole pieces during operation is desirable. (fig. 14).

Great care was taken to minimize the spherical aberration of the objective lens ($C_s \approx 0.5$ mm)[6])[7]).

At first the secondary image was observed through a $10 \times$ binocular. Later on it was found that a $30 \times$ magnification gives much better results.

§ 11. *Focusing of the secondary image.*

Due to image errors the focusing indication becomes less pronounced. Nevertheless after some exercise it is possible to get a reproduceable adjustment. Fig. 15 gives two photographs taken one after the other, such that the second is taken after de- and refocusing. By using auxiliary a-c fields, which can be introduced rather simply, it is possible also for untrained hands to increase strongly the reproduceability of the adjustment. We therefore apply two magnetic 50 c/s fields perpendicular to the optical axis and perpendicular to one another having 90° phase difference. The primary image generally describes an ellipse, the axes of which are functions of the fields intensities. It can be shown easily that the secondary image also gives an ellipse. The vital point is to make one axis of the ellipse in the secondary image as small as possible while keeping the dark centre visible. Of course the minimum value of this field depends on the sharpness of the primary image. Consequently if this image is sharper the short axis of the ellipse can be made smaller before the ellipse collapses into one line. Strictly speaking we determine the resolving power for two lines. We can choose the size of the long axis freely. For a small δ with a very clear secondary image we can decrease the brightness by taking the long axis large. Moreover, if the field can be rotated, which can be done easily electrically, astigmatism in the image can be detected. This auxiliary field can also detect 50 c/s variations in the anode voltages and lens currents.

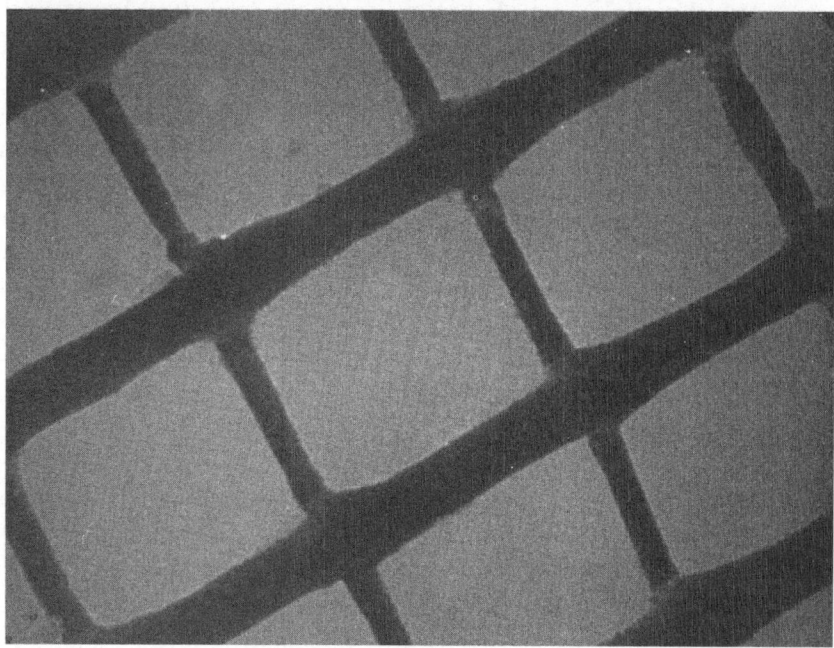

Fig. 15.
Two pictures of 1500 mesh per inch silver grid demonstrating the focusing accuracy. The lower figure is taken after defocusing and refocusing. Au target 6 kV, 20 mins exposure, camera length 10 mm, magn. ca 2000 X.

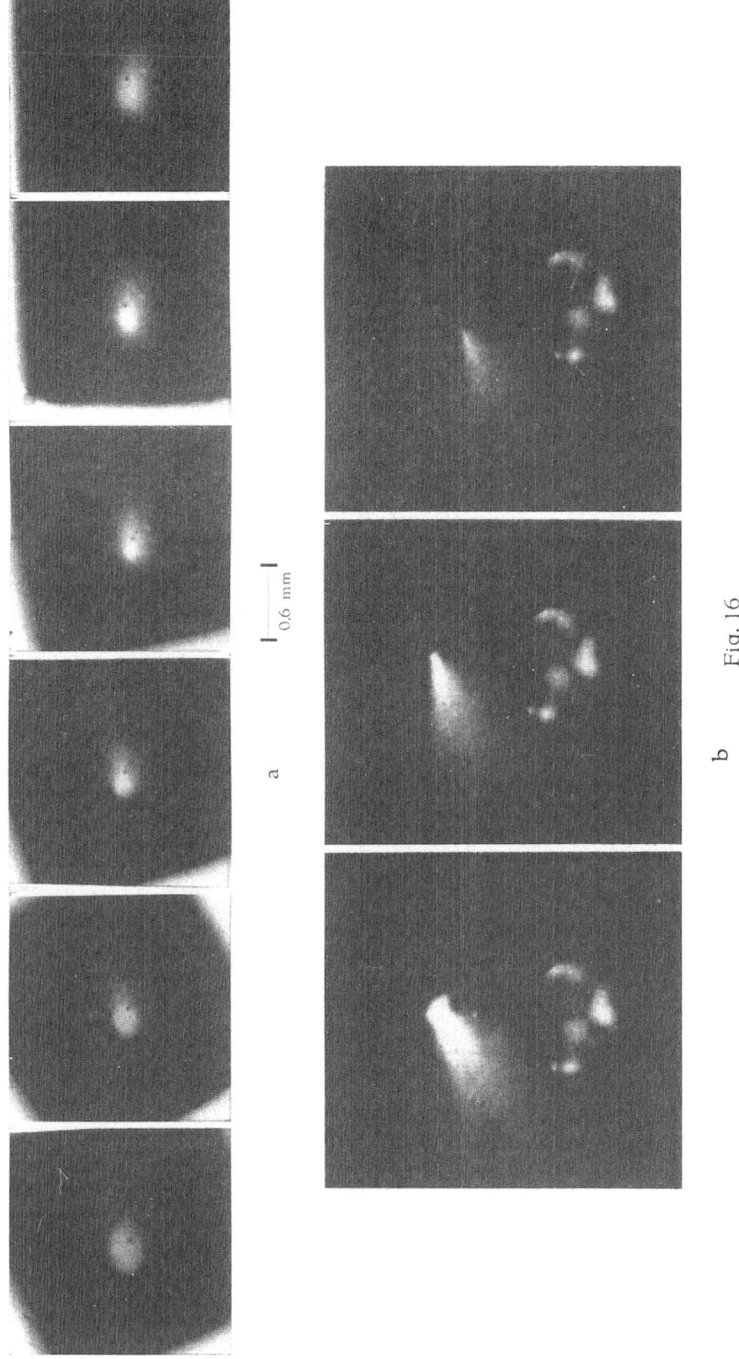

a

0.6 mm

b

Fig. 16

Through-focal series of the secondary image photographed directly from the focusing screen. Increasing lens power from left to right. Upper figure, instrument
aligned; lower figure, aperture not well centred.

§ 12. *Experimental results.*

Centering the condenser according to (2) can be done very simply. Due to the small distance from this lens to the electron source and the small rotation angle θ, however, the accuracy is not high. It is advisable to choose the position of the intermediate target so that $\theta \approx 90°$. In our present apparatus this was not taken into consideration.

To align the objective and centre the aperture stop the projection diagram as described before has proved very valuable. When the microscope is correctly aligned, focusing is achieved very easily since the focusing point is very critical. In fig. 16*a* a through-focal series of the secondary image is shown, taken directly from the screen. From left to right the lens power increases in regular steps. The asymmetry of the focusing curve is obvious. The hole in the screen is faintly distinguishable. The border of this hole is contaminated by fluorescent material which lights up strongly under irradiation by electrons, scattered at the diaphragm below. Fig. 16*b* shows a through-focal series of the secondary image with the lens aperture not correctly centered. Focusing is impossible in this case.

§ 13. *Target contamination.*

It seems that carbon contamination is not directly serious. A target, previously covered with a discharge carbon film of about 500 Å, at first gives a vague reflection image. When the microscope is nearly in focus, the brightness and sharpness increase until after 1 or 2 minutes the final brightness is attained. After electron bombardment during a long period the target roughens. Prolonged high-loading of the same area leads to crater formation. The roughness is visible in the secondary image if the lens is slightly under focus. In this case the apparatus acts as a reflection microscope and vague spots appear. When the microscope is focused, the brightness of the secondary image depends highly on the irradiated part of the target. These effects should be borne in mind in those cases where the X-ray intensity must be constant during long periods. Fig. 17 shows an electron micrograph of a pre-shadowed carbon replica of an irradiated target. The craters caused by intense electron bombardment during long periods are clearly distinguishable. The maximum distance between the craters amounts to some 10 μ. Since the focal length was $\approx$ 1 mm, the maximum tilt

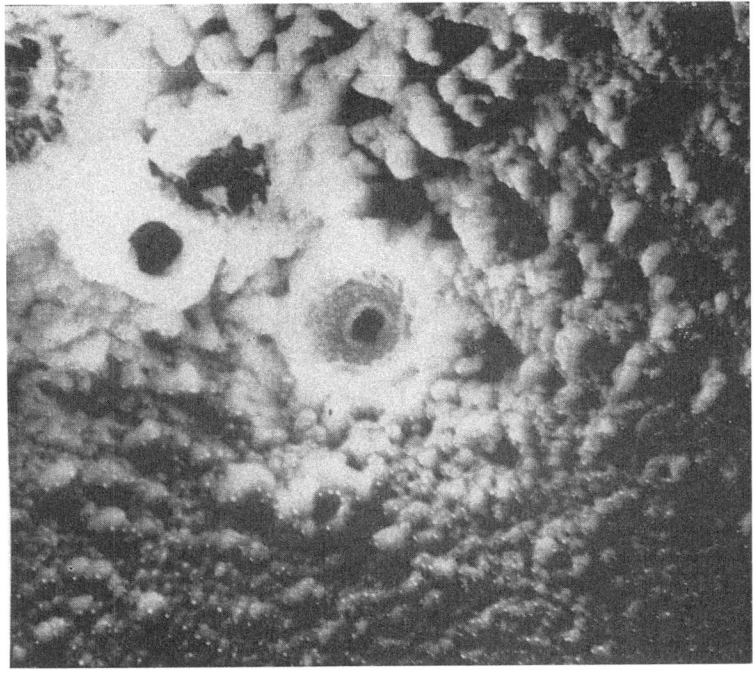

Fig. 17
Electron micrograph of an irradiated target. Pre-shadowed carbon replica. Magnification ca 7000 X. (Courtesy Electron Microscope division of the Technical Physics Department T.N.O. and T.H.)

used was some 1/200 radians. At larger tilt angles the effect shown in fig. 16b appears. Around the craters an area of some 50 μ in diameter is very rough. Fig. 18 shows an electron-micrograph of a part of such an area.

It thus appears that the maximum admissible tilt angle for correct focusing amounts to $\approx$ 1/200 radians. If the target is fixed with respect to the objective, this implies that only an area with radius

$$r = f/200 \qquad (38)$$

can be utilized, f being the focal distance of the objective. In view of the roughening and crater formation it is advisable to make the target movable with respect to the pole-pieces of the lens. If necessary a suitable part of the target could be selected for each exposure. Refocusing, of course, offers no difficulties at all. Finally we show some microradiographs, demonstrating the resolution that

Fig. 18

Electron micrograph of the rough area around the craters. Pre-shadowed carbon replica. Magnification ca 7000 X. (Courtesy Electron Microscope division of the Technical Physics Department T.N.O. and T.H.)

has been obtained. The anode voltage used was always 6 kV. At this voltage the depth of penetration of the electrons in gold according to the Thomson-Whiddington law[12]) amounts to approximately 500 Å. So for a 0.1 μ resolution we do not need an unsupported target of 0.1 μ thickness as Nixon[8])[9]) does. The target we use is an evaporated 5 μ Al-foil, coated with a gold layer of 0.1 μ. The exposure time is some 20 minutes at a film distance of 1 cm. The resolution seems to be limited by insufficient stability (electrical, mechanical and thermal) during long exposure. Fig. 19 shows a microradiographs of gold-shadowed bull sperms. The heads are approximately 10 μ long. The tails show details reminiscent of electron micrographs. Fig. 20 shows untreated bull sperms. Although the contrast is very poor, particularly in the flat head, the first fringe is still visible.

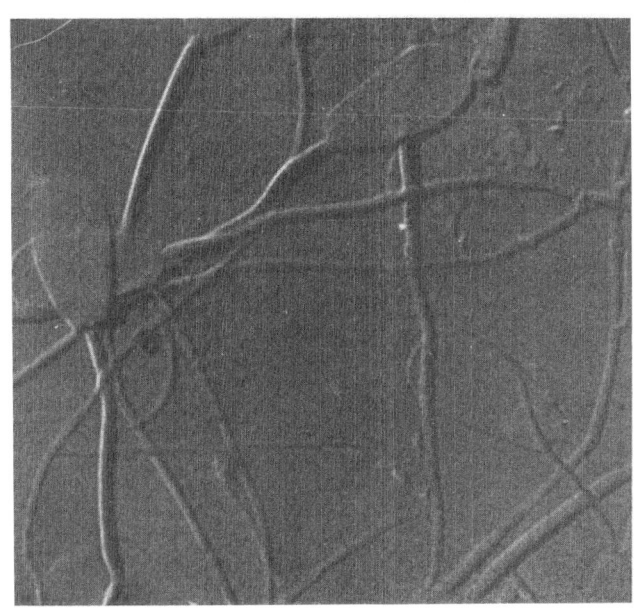

Fig. 19
Gold shadowed bull sperms, magnification ca 2000 X, exposure condition as in fig. 15.

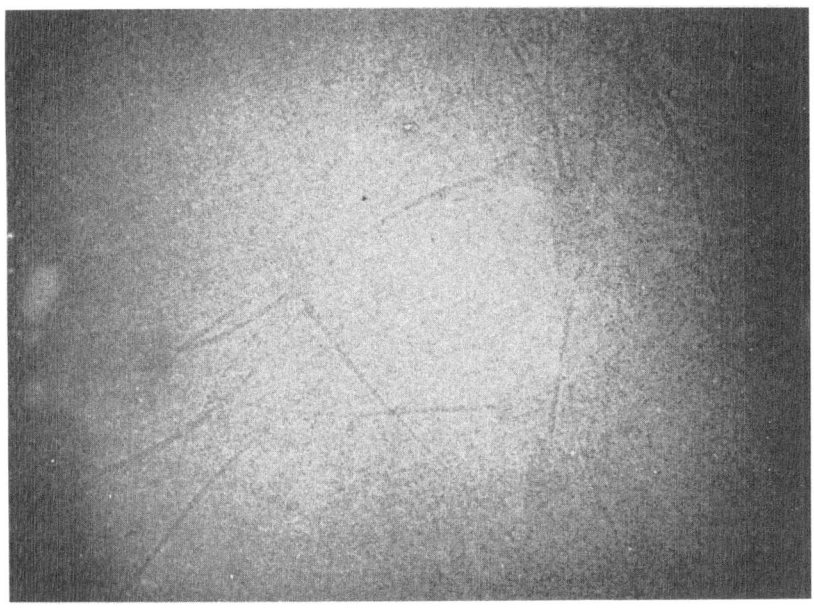

Fig. 20
Untreated bull sperms, magnification ca 1300 X, exposure condition as in fig. 15.

The main part of this chapter has been previously published in
Appl. Sci. Res. *B 7* (1958) 233.

REFERENCES

1) Ardenne, M. v.   Naturwissenschaften **27,** (1939) 485.
2) Ardenne, M. v.   Tabellen der Elektronenphysik Band I, Deutscher Verlag der Wissenschaften, Berlin, 1956, p. 109.
3) Bruining, H.   Physics and applications of secondary electron emission. Pergamon Press Ltd, London 1954.
4) Cosslett, V. E. and W. C. Nixon   Proc. Roy. Soc. **B140,** (1952) 422.
5) Jonker, J. L. H.   Phil. Res. Rep. **6,** (1951) 372.
6) Liebmann, G. and E. M. Grad   Proc. Phys. Soc. **64B** (1951) 956.
7) Ments, M. v. and J. B. Le Poole   App. Sci. Res. **B1** (1947) 1.
8) Nixon, W. C.   Proc. Roy. Soc. **A 232,** (1955) 475.
9) Nixon, W. C.   Nature **175,** (1955) 1078.
10) Rudberg, E.   Proc. Roy. Soc. **127 A** (1930) 111.
11) Rudberg, E.   Phys. Rev. **50,** (1936) 138.
12) Whiddington, R.   Proc. Roy. Soc. **89 A** (1914) 554.

CHAPTER IV

FILM MATERIALS

§ 1. *Introduction.*

The types of film we use for the registration of X-ray micro-scopical images can be divided roughly into three classes:

a) Special, coarse-grained X-ray film, which is exclusively used for recording hard X-rays. (Anode voltage more than 40 kV) Its characteristic features are: high sensitivity and large grains.

b) Normal fine grain film, preferably unsensitised. This is the film used amongst other things for recording electron images, and for making transparencies. Specific features are good contrast and relatively small grains; the sensitivity is considerably lower than a).

c) Ultra fine grain film of the Lippmann type, with submicro-scopical grains (≈ 500 Å). Up till now this class of film is the only one which can be used for high resolution contact microradiography.

Unlike a), the types of film b) and c) are not specially intended for recording X-ray images. The requirement of good resolving power leads to the use of ultra fine grain film with the contact method. With the projection method the resolving power of the film plays a minor part as the primary magnification can always be adapted to the kind of film. Our preference for the normal fine grain film (type b) over the X-ray film (type a) can be accounted for as follows: Film of type a) is manufactured specially for hard X-rays. It has no special advantages when using the relatively soft X-rays that are desirable for microscopy. The advantage of fine grain film lies in its ability to record more information per unit area, so that storing the film takes less room. For printing at about 10 times magnification, an enlarger belonging to the normal laboratory equipment, can be used.

In this chapter we shall examine how far we have utilized the possibilities of the film material. Furthermore we shall try to find an expression for the quality of a film for projection X-ray microscopy.

§ 2. The density curve for X-rays.

The two most important properties of the film for X-rays which differ completely from those for light, are:

1) Practically every absorbed X-ray quantum makes one or more grains developable.

2) Up to a certain value of the density there is a strictly linear relation between density and number of absorbed quanta.

The first statement does not hold for extremely low energies. The second property is an immediate consequence of the first one. According to Nutting [12]) we may write for the transmission of a negative.

$$T = (1 - A\,n)^m \approx e^{-m\,n\,A} \tag{1}$$

in which A is the area of a silver grain, n the number of grains per unit area per layer, and m the number of layers with a thickness equal to the grain diameter of which the emulsion is supposed to consist. For the density,

$$D = \frac{m\,n\,A}{2.3} = \frac{NA}{2.3} \tag{2}$$

in which $N = mn$ equals the total number of grains per unit area, and this in turn is proportional to the number of X-ray quanta per unit area. Recent measurements of Engström and Lindström [6]) show that the ultra fine grain films show linearity up to $D \approx 1.4$.

§ 3. Film data.

Some concepts of photography, i.e. resolving power, sensitivity and contrast, are often introduced in microradiography to characterize the film. The desirability to define other data, completely adapted to projection X-ray microscopy, will be shown.

The resolving power of a film is normally defined as the number of lines per mm that can be separated at a complete black-white transition. The corresponding resolved distance δ is apparently proportional to a, the grain diameter. As a normal image consists of half tones this definition of the resolving power is not of direct practical importance, either for X-rays or for light.

The sensitivity for X-rays is defined as being inversely proportional to the number of incident quanta per unit area necessary to cause a certain density.
So

$$S \infty \frac{D}{N} \qquad (3)$$

in which S is the sensitivity.
From equation (2) it follows that $S \infty A$, the grain area.
Hence

$$\frac{S}{\delta^2} = c \qquad (4)$$

in which c is a constant, determined only by the absorption of the film. From this it would follow that every kind of film, provided it absorbs enough radiation, is equally suitable for use in the projection microscope, since with a small value of δ the primary magnification could be reduced and thus the film be brought proportionally closer to the source. As the number of quanta per unit area on the film increases quadratically, and the sensitivity decreases in the same way with δ according to (4), the total exposure time for getting a given density would remain constant. This generally prevailing conception has caused little attention to be paid to the selection of suitable films for projection microscopy. The slope γ of the film would be decisive in the choice of the film. To obtain a high value of γ a strong developer is used. The slope is, per definition

$$\gamma = \frac{dD}{d \log E} \infty \frac{dD}{d \log X} \infty \frac{dD}{d \log N} \qquad (5)$$

in which E is the relative exposure (for light), X the number of incident quanta (for X-rays) per unit area and N the number of grains per unit area. Substitution of (2) in (5), however, gives

$$\gamma \infty AN \infty D \qquad (6)$$

Whereas for light γ is rather well defined, it has no significance for X-rays as the density varies widely over the picture, and so consequently does γ.

If we want to try to define some practical film data the question may arise as to which factors limit the film in its usefulness. In other words: if we want to record a quantum density image we must study the noise phenomena which may influence this image unfavourably. This will be done in the following §§.

§ 4. *Film noise.*

By film noise we understand the statistical density fluctuations occurring in an originally evenly exposed film, the causes of which are as follows:

1) The density of a film is the result of absorption and scattering by discrete grains.
2) The grains are not homogeneously distributed.
3) The grains are unequal in shape and size.

Film noise has been the subject of extensive research for a long time. Depending upon the definitions and the measuring methods applied, the results, however, are different. (For a summary of the various definitions and measuring methods see Jones and Higgins[9],[10])). Thus in the subjective graininess, determined with the apparatus of Jones and Deisch[1]) a maximum occurs at a density of 0.2 to 0.4. If however the field brightness is kept constant the graininess is directly proportional to the density[11]). Selwyn's[15]) definition of the objective granularity is the most suitable one from our viewpoint. Selwyn based his definition on the probability of the appearance of an image element (with size of δ^2), with a density deviation ΔD with respect to the average density D. In the interval ΔD, d ΔD this probability will be

$$P(\Delta D, \mathrm{d}\Delta D) = \sqrt{\frac{\delta^2}{\pi\, G^2}}\; e^{-\frac{\Delta^2 \delta^2}{G^2}}\; \mathrm{d}\,\Delta D \tag{7}$$

in which G is defined as the granularity constant. Apparently the relation between the standard deviation σ_D of the density fluctuation and G is

$$G = \sigma_D\, \delta\, \sqrt{2} \tag{8}$$

Siedentopf[18]) however, succeeded in relating σ_D to a more fundamental film parameter, i.e. the standard deviation of the grain size distribution σ_A. If the standard deviation of the number of grains in an image element is σ_ν, and that of the average grain area (averaged over an image element) $\sigma_{\overline{A}\nu}$; if furthermore per image element there are on an average ν grains, and if $\overline{A}$ is the mean value of the grain area (averaged over all grains), then according to Siedentopf

$$\frac{\sigma_D}{D} = \sqrt{\left(\frac{\sigma_\nu}{\nu}\right)^2 + \left(\frac{\sigma_{\overline{A}\nu}}{\overline{A}}\right)^2} \tag{9}$$

Denoting the standard deviation of the grain distribution curve by σ_A' then

$$\sigma_{\overline{A}\nu} = \frac{\sigma_A'}{\sqrt{\nu}} \tag{10}$$

For a completely statistical distribution of the grains in the film

$$\sigma_\nu = \sqrt{\nu} \tag{11}$$

This gives for (9)

$$\frac{\sigma_D}{D} = \sqrt{\frac{1}{\nu}} \ \sqrt{1 + \left(\frac{\sigma_A'}{\overline{A}}\right)^2} \tag{12}$$

Siedentopf verified the validity of equation (11) by counting the grains while using light rays. We want here to introduce the dimensionless magnitude

$$\sigma_A = \frac{\sigma_A'}{\overline{A}} \tag{13}$$

and shall understand by "film noise"

$$\frac{\Delta D}{D} = \frac{\sigma_D}{D} = \sqrt{\frac{1}{\nu}(1 + \sigma_A^2)} \tag{14}$$

So according to Siedentopf it does not make a fundamental difference whether the noise is determined for a large image element with a low density or for a small image element with a high density provided in both cases the number of grains is the same. In his derivation Siedentopf started from the validity of equation (9). It must be pointed out however, that equation (9) is strictly speaking only true if there is no correlation between the number of grains in a certain image element and the average value of the grain area of the image element concerned. This only holds if the density is not too small. If for (14) we write

$$\frac{\Delta D}{D} = \sqrt{\frac{1 + \sigma_A^2}{N \delta^2}} \tag{15}$$

with (2) this gives

$$\frac{\Delta D}{D} = \sqrt{\frac{\overline{A}(1 + \sigma_A^2)}{2.3\, D\, \delta^2}}$$

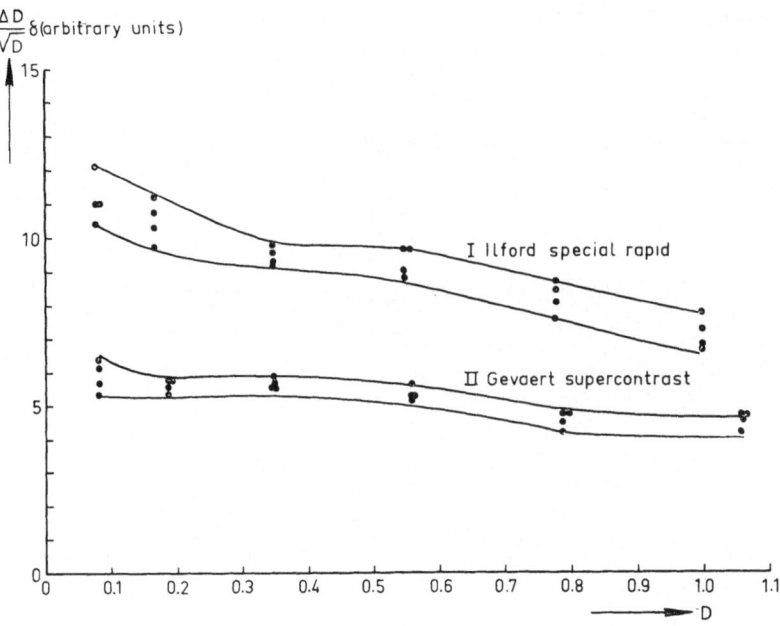

fig. 1

The function $\Delta D\ \delta/\sqrt{D}$, computed from data published by Scheffer [14]).

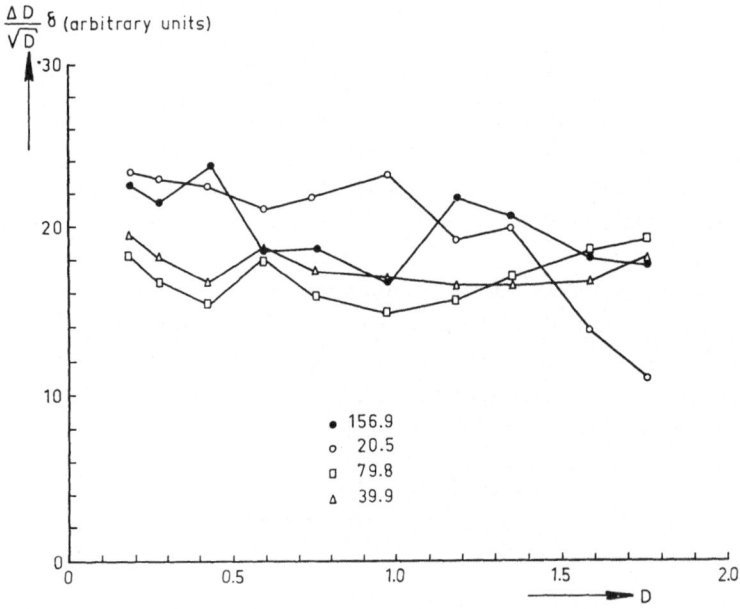

fig. 2

The function $\Delta D\ \delta/\sqrt{D}$, computed from data published by Jones and Higgins [10]).

or
$$\frac{\Delta D \delta}{\sqrt{D}} = \sqrt{\frac{A (1 + \sigma_A{}^2)}{2.3}}. \tag{16}$$

As at small densities the occurence of the larger grains prevails [7, 17]) it may be expected that the function $\Delta D \delta/\sqrt{D}$ decreases with increasing density. The function

$$\sqrt{\bar{A}} = F_{(D)} \tag{17}$$

can be determined by plotting out $\Delta D \delta/\sqrt{D}$ as a function of D. Using results published by Scheffer[14]) and Jones and Higgins[10]) the author has derived the diagrams in fig. 1 and 2 showing the expected deviation. However, as the deviations are rather small, in practice the validity of equation (9) can be assumed with a certain reservation.

§ 5. *Quantum noise of the X-ray image.*

Due to the quantum nature of X-rays, the image will show quantum noise. Sturm and Morgan[19]) proved by means of statistical considerations that this will limit the perceptibility of detail. An image element formed by ξ quanta will show fluctuations with the standard deviation:

$$\sigma_x = \sqrt{\xi} \tag{18}$$

Two image elements formed by ξ and $\xi + \Delta\xi$ quanta respectively can be considered as caused by different absorption in the object with reasonable certainty only if

$$\Delta\xi \geqslant p\sqrt{\xi} \tag{19}$$

Here p is a practical constant, equal to about 5 (according to Tol and Oosterkamp[20]) p may be 3).

At the successive quantum processes (absorption, transformation, intensification) the relative value of the noise will generally increase; it can at best remain the same. As up till now in X-ray microscopy image transfer always takes place with the aid of a photographic plate, it is important to know how the image is influenced by the film noise. The quanta undergo the following processes:

1) *Absorption.* Only a fraction β of the incident quanta is absorbed by the emulsion. The quantum noise is then determined by the number of absorbed quanta.

2) *Transformation.* An absorbed quantum may give rise to a certain number of silver atoms.

3) *Amplification.* By means of the developing process the latent image will be made visible, a process comparable with amplification. The "amplification factor", however, is proportional to the AgBr area.

So the most important factor is not the value of the amplification factor, but the standard deviation in this quantity for the different grains. This is expressed by the parameter σ_A. At a density D the number of silver grains per image element will amount to

$$\nu = \eta \xi \beta \tag{20}$$

in which η is the quantum yield, i.e. the number of grains per absorbed X-ray quantum, ξ the number of incident X-ray quanta per image element and β the absorption factor of the emulsion. If there is no correlation between η and $\xi \beta$ it can easily be shown that

$$\left(\frac{\sigma_\nu}{\nu} \right)^2 = \left(\frac{\sigma_{\overline{\eta}}}{\overline{\eta}} \right)^2 + \left(\frac{\sigma_{\xi\beta}}{\xi\beta} \right)^2 \tag{21}$$

in which σ_ν, σ_η and $\sigma_{\xi\beta}$ are the standard deviation in ν, in the mean value of η and in $\xi \beta$ respectively. As η is an average over $\xi \beta$ quanta, we may write

$$\sigma_{\overline{\eta}} = \frac{\sigma_\eta{}'}{\sqrt{\xi\beta}} = \frac{\overline{\eta}\, \sigma_\eta}{\sqrt{\xi\beta}} \tag{22}$$

in which $\sigma_\eta{}'$ is the standard deviation in η, and σ_η the relative value of this figure.

With $\sigma_{\xi\beta} = \sqrt{\xi\beta}$ equation (21) will be

$$\frac{\sigma_\nu}{\nu} = \sqrt{\frac{\sigma^2{}_\eta + 1}{\xi\beta}} = \sqrt{\frac{\eta}{\nu}(\sigma^2{}_\eta + 1)} \tag{23}$$

When exposing with X-rays we may write for the film noise (by substituting (20) and (23) in (9))

$$\left(\frac{\Delta D}{D} \right)_x = \frac{1}{\sqrt{\xi\beta}} \sqrt{1 + \sigma^2{}_\eta + \frac{\sigma^2{}_A}{\eta}} \tag{24}$$

Concerning the quantum yield η the following should be noticed.

Eggert and Noddack [2, 3, 4]) found a value of $\eta \approx 1$ for $\lambda = 0,45\,\text{Å}$ for different kinds of film (Agfa Röntgenfilm, Agfa Zahnfilm and Agfa Kinepositivfilm). For high energy particles this[5]) amounts to

66

$\eta = 5$ to 15. Measurements of Engström and Lindström[6]) show a linear relation between the number of incident quanta and the density when using ultra fine grain film ($\lambda \approx 3$ Å). This points to η being $\geqslant 1$ for these kinds of film also. Henke[8]) however shows that for the ultra soft rays ($\lambda = 23{,}3$ Å) the density varies logarithmically with the number of incident quanta. This however need not be the direct consequence of the fact that $\eta < 1$. In § 8 of this chapter it will be shown that as the radiation becomes softer the linear relation of the density curve holds only for small density values.

Equation (24) holds under the reservation that there is no correlation between η and $\xi\beta$. For light rays this is not satisfied because more light quanta are needed for developing one grain. This is demonstrated in the density curve varying logarithmically with the number of quanta incident per unit area (X_l). As $\eta \propto \dfrac{D}{X_l}$ the quantum yield for light may be expressed by $\eta_l \propto \dfrac{\log X_l}{X_l}$. For X-rays there is a linear relation between D and $X\beta$ up to a certain value of D. As long as this is the case η is a constant and equation (24) can be used. The graininess of the film for different densities, exposed to X-rays and light is shown in fig. 3.

§ 6. The quality of a film.

In connection with the consideration given in § 4 and § 5 we shall now define a quality measure for the film in projection microradiography. Consider first equation (24), which can be written according to (20) as

$$\frac{\Delta D}{D} = \frac{1}{\sqrt{\nu}} \sqrt{\eta (1 + \sigma^2{}_\eta) + \sigma^2{}_A} \qquad (25)$$

and compare this with equation (4) for light rays.

For $\eta (1 + \sigma^2{}_\eta) > \sigma^2{}_A$ and $\eta (1 + \sigma^2{}_\eta) > 1$ the negative for X-rays at equal density will look more "granular" than for light rays; in this case the film also records the X-ray quantum noise, which would point to a film of good quality. If however $\eta(1+\sigma^2{}_\eta) \approx 1$ and $\sigma^2{}_A > 1$ the density fluctuation will mainly be film noise. In general information will be lost.

The fact that quantum noise can be shown in the negative does

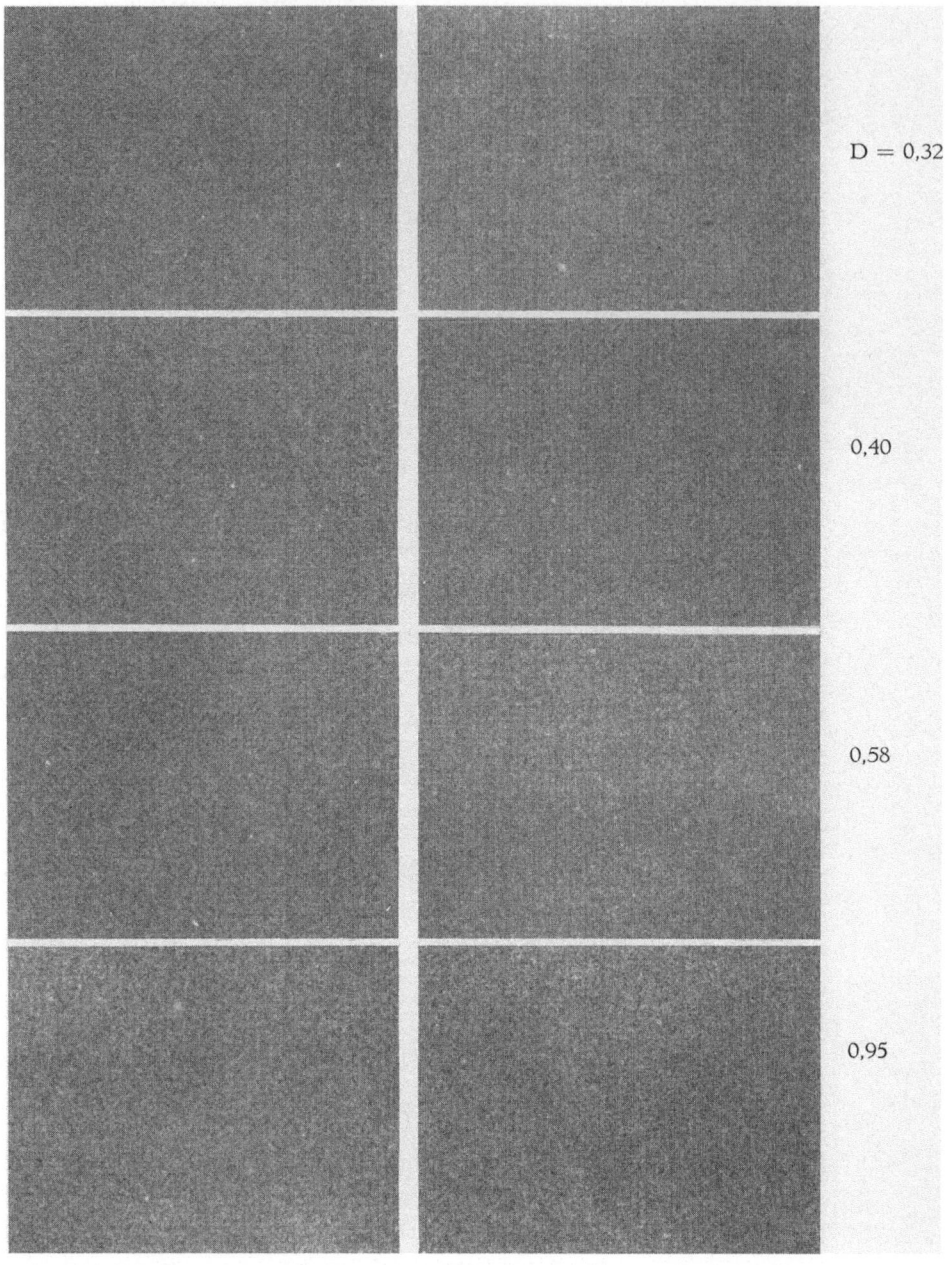

D = 0,32

0,40

0,58

0,95

fig. 3
Graininess of Ilford Process film for light (right) and X-rays (left) at different
densities (30 X)

not yet guarantee that the utmost is attained. This holds only if $\beta = 1$. For projection microscopy a film of the highest quality should have such properties that the image noise in the negative is equal to that of the original X-ray image. If we indicate the film quality by q we could put

$$\frac{1}{q} \ \infty \ \frac{\left(\dfrac{\varDelta D}{D}\right)_x}{\dfrac{\sigma_\xi}{\xi}} \tag{26}$$

With (24) and (18) we get

$$\frac{q_o}{q} = \frac{\left(\dfrac{\varDelta D}{D}\right)_x}{\dfrac{\sigma_\xi}{\xi}} = \frac{1}{\sqrt{\beta}} \ \sqrt{1 + \sigma^2{}_\eta + \frac{\sigma^2{}_A}{\eta}} \tag{27}$$

in which q_o is an arbitrarily definable figure. This would mean however, that a film with a k times larger absorption is only $\sqrt{k}$ times better. To get equal results the exposure time for the poorer film must be k times longer. As the exposure time plays such an important part, especially for projection microscopy, we prefer to define the quality by

$$Q = Q_o \ \frac{\left(\dfrac{\sigma_\xi}{\xi}\right)^2}{\left(\dfrac{\varDelta D}{D}\right)^2} = \frac{Q_o \ \beta}{1 + \sigma^2{}_\eta + \dfrac{\sigma^2{}_A}{\eta}} \tag{28}$$

§ 7. *Visual quality comparison of a film.*

The quality of a film, as defined in equation (28), can only be determined with specially constructed apparatus. Besides, at this stage we are not so much interested in expressing the quality as a numerical value as in the comparison of existing kinds of film. In this paragraph a method will be described schematically, which can

be used for this purpose without special apparatus. If $\frac{\sigma_\xi}{\xi}$ in equation (28) is kept constant the quality is inversely proportional to $\left(\dfrac{\varDelta D}{D}\right)$. By visual comparison of the graininess an impression of the film quality can be obtained, at least we can determine the quality in comparison with a given film. For this purpose the negative is magnified and printed, and the noise images are compared. We must take care, however, that the gradation γ of the paper is the same for the different negatives, so the developing conditions and the kind of paper must be constant. Besides, for equal γ the value of $\frac{\varDelta D}{D}$ of negative and print are not equal, so that only negatives with equal densities may be compared. In general $\frac{\sigma_\xi}{\xi}$ is also different for the various kinds of film. The magnification however can be chosen such that $\frac{\sigma_\xi}{\xi}$ or at least ξ is the same for every print. Suppose the quantum density of the incident radiation for obtaining a density D amounts to X_1, X_2 etc., for the various kinds of film. Here $X \propto It$, in which I is the radiation intensity and t the exposure time (see fig. 4). For an optical magnification M_0 the corresponding quantum density on the print amounts to

$$X_p \propto \frac{X}{M_o^2}.$$ (29)

If we want to keep its value constant for the different prints

$$\frac{X_1}{M_1^2} = \frac{X_2}{M_2^2} = \text{etc.} \qquad \text{must be constant} \qquad (30)$$

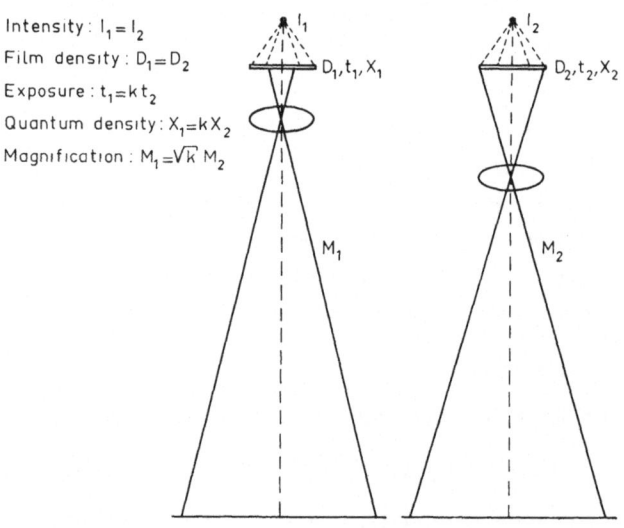

Intensity : $I_1 = I_2$
Film density : $D_1 = D_2$
Exposure : $t_1 = k t_2$
Quantum density : $X_1 = k X_2$
Magnification : $M_1 = \sqrt{k}\, M_2$

fig. 4
Condition for visual quality comparison of the film.

If the radiation intensity is kept constant when exposing, the following condition must be satisfied.

$$M_o \propto \sqrt{t} \qquad (31)$$

Instead of varying the exposure time with constant X-ray intensity and constant film to source distance we can vary the latter and keep the exposure time constant. Although this method corresponds more to reality (in projection microscopy), in practice it is difficult to realize. In the first procedure the film to source distance need not be known. From a series of exposures the parts with the correct density are selected and printed on the right scale. It is necessary to check if the position of the chosen value of D is in the linear part of the density curve.

§ 8. *The resolving power of a film.*

We consider two adjacent image elements of a film with densities D_1 and D_2. We can only say with reasonable certainty that they are conjugate to two parts of the object with different absorption if

$$\frac{D_1 - D_2}{D_1} \geqslant 5 \left(\frac{\Delta D}{D} \right)_x \qquad (32)$$

in which $\left(\dfrac{\Delta D}{D} \right)_x$ represents the noise in the film image.
In the X-ray image

$$\frac{X_1 - X_2}{X_1} = C \qquad (33)$$

represents the image contrast. In the linear part of the density curve also

$$\frac{D_1 - D_2}{D_1} = C \qquad (34)$$

Substitution of (25) and (34) in (32) with $\nu = N\delta^2$ gives after some transformations

$$\delta \geqslant \frac{5}{C\sqrt{N}} \sqrt{\eta(1 + \sigma^2_\eta) + \sigma^2_A} \qquad (35)$$

This equation represents the minimum condition that must be satisfied to consider two image elements of a film with reasonable certainty as such. δ is the resolvable distance, the value of which is, as appears from (35) inversely proportional to the image contrast.

This fact was pointed out by Le Poole some years ago.

It seems strange that according to (35) the resolving power does not depend on the grain size, which is of course the case as N cannot be increased without limit. Some conditions concerning the grain area must be added to equation (35).

We consider the film as consisting of m layers with a thickness equal to the grain diameter. (The grains are considered to be small spheres). The maximum number of grains per layer per unit area (see equation (1)) is

$$n_{max} = \frac{1}{A} \tag{36}$$

The linear relation between density and number of incident quanta only holds provided that per layer, only

$$n = \varepsilon\, n_{max} = \frac{\varepsilon}{A} \tag{37}$$

grains are exposed, in which

$$\varepsilon \ll 1 \tag{38}$$

Equation (35) can be written, since $N = mn$, as

$$\delta \geqslant \frac{5}{C} \sqrt{\frac{A}{m\varepsilon}} \sqrt{\eta\,(1 + \sigma^2{}_\eta) + \sigma^2{}_A.} \tag{39}$$

Here ε is a measure of the total number of grains that may be exposed per layer. If $\varepsilon \approx 1$, a large number of grains is "double-exposed", the quantum efficiency η decreases. No linear relation exists anymore between density and number of X-ray quanta absorbed per unit area. In § 5 attention is drawn to the fact that the deduced equation for $\left(\dfrac{\varDelta D}{D}\right)_x$ does not hold any more.

Introducing the density in equ. (35) gives, with (2)

$$\delta \geqslant \frac{1}{C} \sqrt{\frac{10A}{D}} \sqrt{\eta\,(1 + \sigma^2{}_\eta) + \sigma^2{}_A.} \tag{40}$$

The advantage of this equation is the fact that we can get an impression of the maximum possible resolving power by determining the density curve, which is important for contact microscopy. The smallest possible resolvable distance, for a certain contrast C, is obtained by substituting $D = D_u$ in equation (40), in which D_u (useful density) is the maximum value of D for which the deviation from linearity has a just admissible value.

72

Suppose the grain diameter of the film is a, the absorption coefficient μ, and the number of X-ray quanta incident per unit area X. Then the total number of quanta per unit area, absorbed in the first layer of the film, amounts to

$$X_1 = X\,(1 - e^{-\mu a}) \qquad (41)$$

This gives rise to the formation of

$$n_1 = \eta X\,(1 - e^{-\mu a}) \qquad (42)$$

silver grains. To prevent double exposure every layer must satisfy equations (37) and (38), so that for the first one, which is the most exposed, we have

$$\frac{\varepsilon}{A} \geqslant \eta X (1 - e^{-\mu a}) \qquad (43)$$

For exposure to a density D_u we may equate both sides of equation (43), and the number of incident X-ray quanta allowed to hit the film per unit area can be computed. Hence

$$X_{max} = \frac{\varepsilon}{A\,\eta\,(1 - e^{-\mu a})} \qquad (44)$$

For total absorption of the radiation in the film the number of grains per unit area is

$$N_{max} = \frac{\varepsilon}{A\,(1 - e^{-\mu a})} \qquad (45)$$

so that with (2)

$$D_u = \frac{\varepsilon}{2.3\,(1 - e^{-\mu a})} \qquad (46)$$

If $\mu a < 1$ we may write

$$D_u \approx \frac{\varepsilon}{2.3\,\mu a} \qquad (47)$$

As $a \infty \sqrt{A}$ from (40) and (47) it follows for small values of μa that

$$\delta \infty a^{3/2} \qquad (48)$$

This result is rather unexpected as up to now a linear relation between δ and the grain diameter a (see § 3) has been generally assumed.

§ 9. *Information transfer with the aid of a print.*

The X-ray image we get from an object consists of quantum density variations which by some transformation we want to carry over
as a signal to our eye. As the film up till now has the best registration qualities, this transformation takes place via the film. The
quantum density variation is converted into a grain density variation where this medium itself introduces noise. The next transformation normally used is the conversion of the original density
variation into a brightness variation with the aid of a print. It
will be shown that, due to the logarithmic response of the photographic process, a good deal of detail will be obscured.

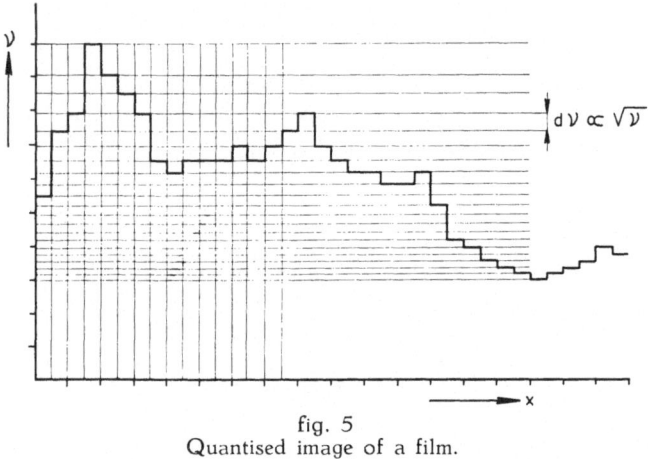

fig. 5
Quantised image of a film.

In fig. 5 the grain density is plotted as a function of the x-coor-
dinate of the film. In the horizontal direction the image is quantised
into line elements with a length of $\Delta x =$ constant; in the vertical
direction into grain numbers of value $d\nu \propto \sqrt{\nu}$. The contribution
which two adjacent image elements make to the total information
is proportional to $\Delta \nu$ divided by $d\nu$. If we call this information
contribution ΔI

$$\Delta I = i \frac{\Delta \nu}{\sqrt{\nu}} \tag{49}$$

in which i is a constant depending on the film noise.

According to the Weber-Fechner law the eye perceives a bright-
ness variation $\dfrac{\Delta B}{B}$ only if that magnitude exceeds a certain value.

Thus the contribution which two adjacent image elements make to the total information, passed on to the eye amounts to

$$\Delta I_e = i_e \frac{\Delta B}{B} \tag{50}$$

in which i_e is constant within very wide brightness limits. If $\Delta I > \Delta I_e$, full information transfer takes place. The eye however perceives the film noise also as (inconvenient) information. For $\Delta I < \Delta I_e$ no information is carried over any more; the eye sees the two image elements equally bright. So it is desirable that

$$\Delta I = \Delta I_e \tag{51}$$

When printing the negative this is not satisfied, as will be shown below. In most cases the density D_p of a printing paper is a logarithmic function of the exposure E. The exposure E is proportional to the transmission of the negative $T = 10^{-NA/2.3}$. For D_p we may write $D_p = \gamma \log C_1 T$

or
$$D_p = - \gamma NA/2.3 + \gamma \log C_1 \tag{52}$$

The brightness variation of the print then amounts to

$$B = B_0 [10 \exp. (\gamma NA/2.3 - \gamma \log C_1)]$$

For each image element we may write

$$B = B_0' e^{\gamma \nu A} \tag{53}$$

Hence
$$\Delta I_e = i_e \frac{\Delta B}{B} = i_e \gamma A \Delta \nu$$

The ratio of the received and available information contributions per image element will be

$$\frac{\Delta I_e}{\Delta I} = \frac{i_e}{i} \gamma A \sqrt{\nu} \tag{54}$$

For a photographic emulsion γ is almost constant, as a result of which the print has a tendency to show a large graininess in the bright part and to obscure details in the dark parts. There is only a narrow brightness region where every detail shows to full advantage.

If γ could be made variable we should need to arrange that

$$\gamma \infty \frac{1}{\sqrt{\nu}}$$

or
$$\gamma \infty \frac{1}{\sqrt{N}} \tag{55}$$

75

With $E \infty 10^{-NA}$ and so $N \infty -\log E$ (55) becomes

$$\gamma \infty \frac{1}{\sqrt{-\log E}}$$ (56)

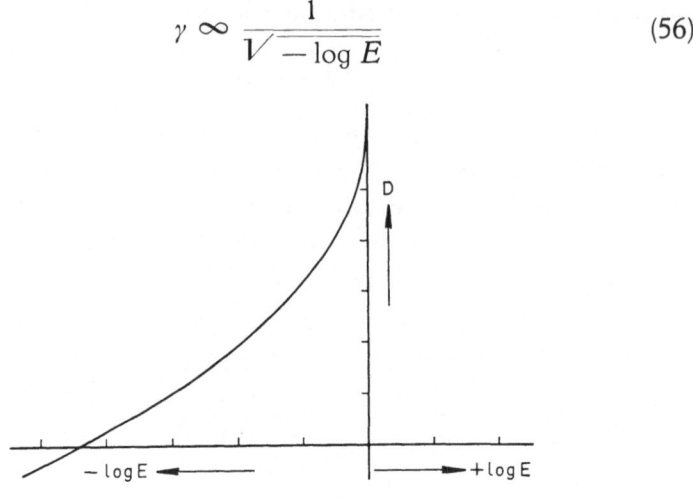

fig. 6

The density curve of a printing paper for correct γ adaptation.

The qualitative behaviour of the density curve satisfying (56) is represented in fig. 6. To approximate such a curve we could make use of the underexposed portion of the density curve of a film. In general the maximum density for which this can be done, and the absolute value of γ, are small. As a result of the inevitable fog, noise is introduced. As densities are additive, Le Poole suggests the approximation of condition (56) by making use of different transparencies, made with different exposure times and different values of γ (fig. 7).

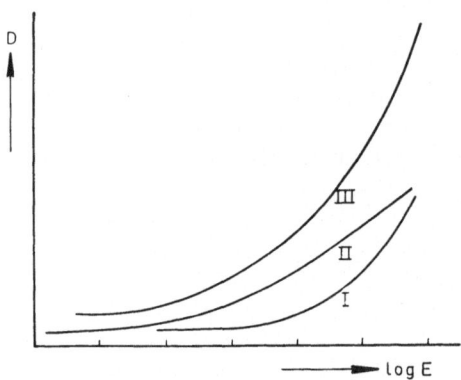

fig. 7

Approximation of the curve in fig. 6 (III) using two transparencies (I and II).

76

Note: The reasoning in § 9 is based on the Weber-Fechner law. This however applies only to an area with a brightness range of some 100 (D_p is maximally 2) [13]). With a transparency a brightness range of more than 1000 can be obtained so that another measuringstaff must be taken. In general we shall have to choose the exposure times and the thickness of the specimen in such a way that a brightness range is obtained which is not too large.

§ 10. *Information transfer with the aid of a fluorescent screen.*

Now that we have considered the information transfer with the aid of a film it is interesting to compare it with the information transfer with the aid of a fluorescent screen. With this system light emitted by the screen is transferred as a signal to our eye, possibly after amplification. As an intermediate step a photographic process can be used where, however, mainly fluorescent light is used. The brightness of the screen is proportional to the number of quanta per unit area per unit of time. Eliminating the accumulation time of the eye or the exposure time we get

$$B_s \infty \; \xi \tag{57}$$

in which B_s is the screen brightness and ξ the total number of X-ray quanta per image element. The information contribution of the screen per image element then becomes

$$\Delta I_s = i_s \frac{\Delta \xi}{\xi} \tag{58}$$

in which i_s depends on the efficiency of the fluorescent screen and on the amplification factor if an image amplifier is used. The information contribution available is according to (49) equal to $i\frac{\Delta v}{Vv} = i\frac{\Delta \xi}{V\xi}$. The ratio of the received and available information contributions then becomes

$$\frac{\Delta I_s}{\Delta I} = \frac{i_s}{i\sqrt{\xi}} \tag{59}$$

For small values of ξ (dark parts) the information is strongly limited by the noise, whereas for the light parts the contrast is too small to observe details. If an image amplifier of special construction is used, it is possible in principle to apply a gradation (γ) correction. To this end i_s has to satisfy

$$i_s \infty \; \sqrt{\xi}. \tag{60}$$

For visual observation, the original image contrast is small in general (short accumulation time of the eye), so gradation correction does not make much sense.

§ 11. *The image contrast.*

X-ray microscopy in general aims at what is called a "contrasty image", by which is meant mostly such an image that a small variation in μz (μ absorption coefficient, z thickness of the specimen) causes clearly visible brightness variations. To obtain this it is considered desirable to use soft X-rays (compare chapter II, § 6). The answer to the question of how far this supposition is correct, will be dealt with in chapter VI. In this paragraph however, it will be demonstrated that by a suitable choice of film material, geometrical arrangement, and exposure time it is possible in principle to attain the same result with relatively hard (monochromatic) X-rays.

We consider again the quantised density image of fig. 5. Whether or not a small change in μz can be detected is determined by whether this will cause a grain density variation $\varDelta \nu > \mathrm{d}\nu$. At a certain value of variation in μz, however, $\varDelta \nu$ will be proportional to ν whereas $\mathrm{d}\nu$ is proportional to $\sqrt{\nu}$. From this consideration it follows that if we can make the product of exposure time and radiation intensity sufficiently large we can in principle make any variation in μz visible, irrespective of what wavelength is used. This property is difficult to verify for the fluorescent image, where by using shorter wavelengths the image contrast (i.e. the ratio of the lightest and darkest parts) will decrease. At a quantum density of

$$X = X_o \, e^{-\mu z} \tag{61}$$

in which X_o is the quantum density in the absence of the object, the screen brightness will amount to

$$B_s = B_{so} \, e^{-\mu z} \tag{62}$$

in which $B_{so} \propto X_o$. The image contrast then amounts to

$$\frac{B_{\max}}{B_{\min}} = e^{-(\mu z)_{\min} \, + \, (\mu z)_{\max}} \tag{63}$$

and thus is independent of X_o. The result is that we see the image poorer in contrast by using radiation with a shorter wavelength, while by increasing the intensity we cannot improve contrast. This of course holds whilst the intensity variation is within the validity of the Weber-Fechner law. In fig. 8a the quantum density distribution is shown for two values of X_o. These figures also represent the brightness distribution of the screen.

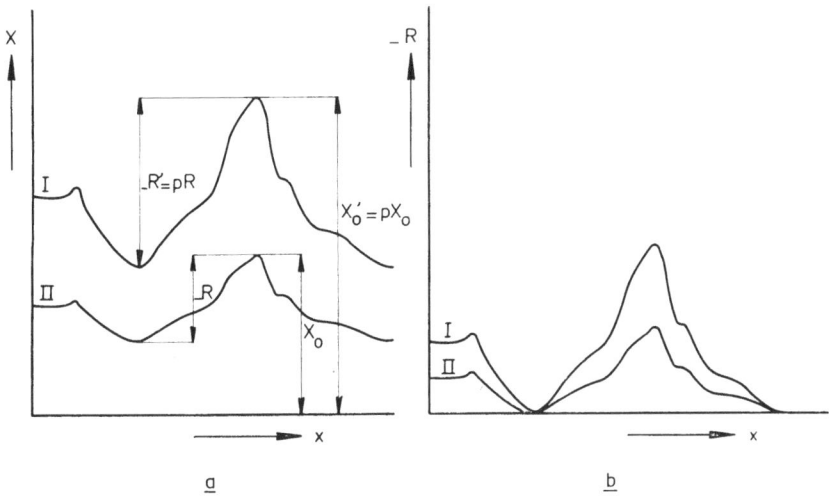

fig. 8
a) X-ray quanta density image for two different exposures.
b) The corresponding values of $-R$ (absorption with opposite sign).

For (61) we write (see fig. 8a)

$$X = X_o - R \tag{64}$$

in which

$$R = X_o (1 - e^{-\mu z}) \tag{65}$$

Here R apparently is the quantum density distribution representing the absorption in the object.

From (65) it follows that R does increase proportionally to X_o. If it is possible to make an image with a brightness proportional to R, a loss in contrast, caused by using hard radiation, can be compensated by a large intensity. In fig. 8b the values of $-R$ deduced from fig. 8a are represented. We could get such an intensity distribution if we could subtract a constant amount, proportional to $X_o e^{-(\mu z)_{max}}$ from the screen brightness. In general we can only do this via an image transformer. For a film, which in fact is an image transformer, the case is more favourable. For the transmission we can write from equation (64)

$$T = T_o e^{-X_o + R} = T_o e^{-X_o} e^{+R}. \tag{66}$$

The transmission apparently is an exponential function of the quantum density, absorbed in the object, and so the image contrast will increase with exposure, a fact which we have already seen in § 3.

At a certain X-ray intensity and film to source distance the exposure time cannot be increased arbitrarily, for the maximum admissible density is determined by the film (c.f. § 8). For longer exposure we must put the film further away from the source. The maximum density variation is now determined by D_u, amongst other things. In itself this presents no serious difficulties as when printing we can use harder paper without the appearance of an inconvenient graininess. Then, however, we can only profit completely by all possibilities if we apply a γ correction as discussed in § 9.

REFERENCES

1) Deisch, H. and L. A. Jones — J. Frank Inst. **190** (1920) 657.

2) Eggert, J. and W. Noddack — Zeit. f. Phys. **43** (1927) 222.

3) Eggert, J. and W. Noddack — Zeit. f. Phys. **44** (1927) 155.

4) Eggert, J. and W. Noddack — Zeit. f. Phys. **51** (1928) 796.

5) Eggert, J. and W. Noddack — Naturwiss. **15** (1927) 57.

6) Engström, A. and B. Lindström — Proc. Roy. Soc. **B 140** (1952) 33.

7) Handbuch der Wissenschaftl. und Angewandten Photographie Band V, Wien, Verlag v. Julius Springer 1932, p. 233.

8) Henke, B. L. — High Resolution Microradiography (1958). Techn. Rep. No 2, Ultrasoft X-ray Physics Air Force Office of Scientific Research, p. 27.

9) Jones, L. A. and G. C. Higgins — J. Opt. Soc. Am. **35** (1945) 435.

10) Jones, L. A. and G. C. Higgins — J. Opt. Soc. Am. **36** (1946) 203.

11) Mees, C. E. K. — Theory of the Photographic Process p. 847. McMillan Company, New York 1945.

12) Nutting, P. G. — Phil. Mag. **26** (1913) 423.

13) Nutting, P. G. — Report of Standards Committee on Visual Sensitometry, J. Opt. Soc. Am. **4** (1920) 55.

14) Scheffer, J. C. — Thesis, Utrecht (1936).

15) Selwyn, E. W. H. — Phot. Journ. **75** (1935) 571.

16) Selwyn, E. W. H. — Phot. Journ. **79** (1939) 513.

17) Sheppard, S. E. A. P. H. Trivelli and E. P. Wightman — J. Frank Inst. **196 II** (1923) 779.

18) Siedentopf, H. — Phys. Zeit. **38** (1937) 454.

19) Sturm, R. E. and R. H. Morgan — Amer. J. Röntgenology Rad. Ther. **62** (1949) 617.

20) Tol, T. and W. J. Oosterkamp — Phil. Tech. Rev. **17** (1955) 65.

CHAPTER V

CONTRAST IMPROVEMENT BY THE USE OF
ULTRA FINE GRAIN FILM

§ 1. *Introduction.*

In chapter V § 1, it is demonstrated that the use of ultra fine grain film is up to now limited to the contact method. The reason that this film is not applied to the projection method is the necessity of subsequent magnification with the aid of an optical microscope. Besides this requiring an extra action, the photographs obtained are rather small. Furthermore it is generally assumed that this kind of film has no particular advantages.

After the work on the focusing method described in chapter III a study on thin biological objects was started in Delft. When using 6 kV anode voltage the contrast however was disappointing (see chapter III fig. 20). Mosley, Scott and Wyckoff[7,8]) obtained excellent results with Mg radiation (10 Å) and 12 kV (contact micrographs). We, therefore, also tried to use this radiation, and although the result was better than when using Au white radiation the contrast was far below that which the cited authors obtained. As the photographs published were contact exposures using ultra fine grain film, it was natural to suppose that this kind of film has better recording properties. A test exposure with Kodak Maximum Resolution gave surprising results. Even if white radiation at 10 kV was used spermatozoa could be made visible. These results necessitated a further study of the properties of photographic emulsions. While chapter IV deals principally with the properties of films under monochromatic irradiation this chapter will deal qualitatively with the effect of non-monochromatic radiation. As will appear, the success of the 2X method[10]) was largely the effect of the film properties we shall also give some further discussion on this method.

§ 2. *The use of non monochromatic radiation.*

Suppose the spectral distribution (fig. 1) of the X-ray source is given by

$$I_\lambda = \Phi(\lambda) \tag{1}$$

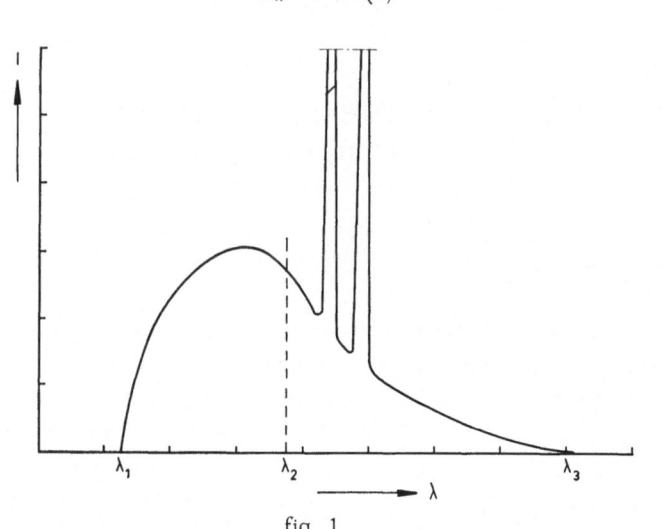

fig. 1

Schematic representation of the spectral distribution of an X-ray source.

in which λ may show discontinuities. Denoting the absorption coefficient of the object by $\mu(\lambda)$ (where $\mu(\lambda)$ is also a function of the coordinates of the object), and the thickness by z we may write

$$\mu(\lambda) z = a_\lambda \tag{2}$$

The intensity distribution of the radiation after passing a given object then amounts to

$$I = \int_{\lambda_1}^{\lambda_3} \Phi(\lambda) e^{-a_\lambda} \, d\lambda \tag{3}$$

We divide the radiation into "hard" and "soft" radiation and consequently write

$$I = \int_{\lambda_1}^{\lambda_2} \Phi(\lambda) e^{-a_\lambda} \, d\lambda + \int_{\lambda_2}^{\lambda_3} \Phi(\lambda) e^{-a_\lambda} \, d\lambda \tag{4}$$

Depending on the function $\Phi(\lambda)$ and the value of a_λ at a given value of λ_2, I_f, as defined by

$$\int_{\lambda_1}^{\lambda_2} \Phi(\lambda)\, e^{-a\lambda}\, \mathrm{d}\lambda \approx \int_{\lambda_1}^{\lambda_2} \Phi(\lambda)\, \mathrm{d}\lambda = I_f \tag{5}$$

will be almost constant over all image elements. Radiation in this part of the spectrum will therefore give practically no information about the object. It does however produce a constant density on the film comparable to fog. Therefore we shall call this part of the radiation "fog-radiation". Similarly for the second term of equation (4) we can write for a given spectrum and a given object

$$\int_{\lambda_2}^{\lambda_3} \Phi(\lambda)\, e^{-a\lambda}\, \mathrm{d}\lambda = I_i\, e^{-a} \tag{6}$$

Here I_i is the intensity of the radiation which we shall call "informatory-radiation", and which is modulated by the wanted information e^{-a} as a function of the coordinates of the object. From the definition of λ_2 it follows that a depends on the spectral distribution I_λ. Equation (4) may now be written as

$$I = I_f + I_i\, e^{-a} \tag{7}$$

Furthermore

$$I_o = I_f + I_i = \int_{\lambda_1}^{\lambda_3} \Phi(\lambda)\, \mathrm{d}\lambda \tag{8}$$

in which I_o represents the intensity of the incident radiation. The intensity distribution along one line on the film is given schematically in fig. 2.

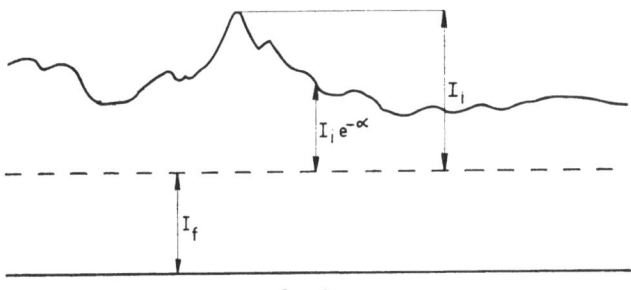

fig. 2

Intensity distribution of the X-ray image at film level.

83

§ 3. The influence of fog-radiation on visual contrast.

The visual contrast of the negative will be defined as

$$\frac{\varDelta T}{T} = \varDelta \ln T \tag{9}$$

for two adjacent image elements in which T is the transmission of the negative. This visual contrast determines the visibility of a detail in the negative and hence in a print later on. It will be shown now that the visual contrast is independent of the fog-radiation as long as the film responds linearly.

For this condition we can write

$$T = e^{k(I_f + I_i e^{-a})} \tag{10}$$

so the visual contrast

$$VC = \varDelta [k(I_f + I_i e^{-a})] = k I_i \varDelta e^{-a} \tag{11}$$

Thus VC is independent of I_f, which by definition is a constant. The fog radiation, however, does influence the maximum value of the density at which a linear relation still exists between the density and the number of incident quanta of informatory-radiation. According to equation (47) chapter IV we may write for the useful density at a given wavelength

$$D_u = \frac{\varepsilon}{2.3 \, \mu_\lambda a} \tag{12}$$

If the absorption coefficient for the informatory radiation is μ_i and for the fog-radiation μ_f we must see to it that the top layer of the film is exposed in such a way that

$$N_i (1 - e^{-\mu_i a}) + N_f (1 - e^{-\mu_f a}) \leq \frac{\varepsilon}{A} \tag{13}$$

Furthermore we put $\quad \dfrac{N_i}{N_f} = C \text{ and } N_i + N_f = N \tag{14}$

Here N is the number of quanta per unit area and the indices i and f refer to the informatory-and fog-radiation respectively. From (13) and (14) we can calculate that for $\mu_i a << 1$ and $\mu_f a << 1$

$$N_i = \frac{C\varepsilon}{Aa \, (C \mu_i + \mu_f)} \tag{15}$$

$$N_f = \frac{\varepsilon}{Aa \, (C \mu_i + \mu_f)} \tag{16}$$

and
$$N = \frac{\varepsilon}{Aa\mu_f}\left(\frac{1+C}{1+C\frac{\mu_i}{\mu_f}}\right) = \frac{\varepsilon}{Aa\mu_i}\left(\frac{1+C}{\frac{\mu_f}{\mu_i}+C}\right) \qquad (17)$$

so that

$$D_u = \frac{\varepsilon}{2.3\,\mu_f a}\left(\frac{1+C}{1+C\frac{\mu_i}{\mu_f}}\right) = \frac{\varepsilon}{2.3\,\mu_i a}\left(\frac{1+C}{\frac{\mu_f}{\mu_i}+C}\right) \qquad (18)$$

for $C \to 0$
$$(D_u)_f \approx \frac{\varepsilon}{2.3\,\mu_f a}, \qquad (19)$$

for $C \to \infty$
$$(D_u)_i \approx \frac{\varepsilon}{2.3\,\mu_i a}. \qquad (20)$$

In general we may write

$$D_u = (D_u)'_i + (D_u)'_f \qquad (21)$$

in which $(D_u)_i'$ and $(D_u)_f'$ are the available density ranges for the informatory- and the fog-radiation respectively.

$$(D_u)'_i \approx C\,(D_u)'_f \qquad (22)$$

$$(D_u)'_i \approx \frac{D_u C}{1+C} \qquad (23)$$

For the normal fine grain film (22) and (23) are rather small for soft radiation. The maximum visual contrast then amounts to

$$\Delta \ln T = -2.3\,\Delta D_u = -2.3\,(D_u)'_i \qquad (24)$$

The visual contrast is influenced by the fog-radiation in that the available density range has become smaller, and so therefore has the visual contrast. For this reason it is desirable to use a very small grain diameter, i.e. ultra fine grain film. Although contrast can be improved in printing, working with a small density range is not desirable because of the inevitable fog, dirt, scratches etc.

§ 4. *The influence of fog-radiation on the film noise.*

As the film noise is a direct function of the absolute density, the fog-radation will introduce noise proportional to $\sqrt{\nu_f}$, in which ν_f is the number of grains per image element, blackend by fogquanta. For the signal to noise ratio, in the presence of fog-radiation, we get

$$\text{SNR} = \frac{v_i\, e^{-a}}{\sqrt{v_f + v_i\, e^{-a}}} = \frac{\sqrt{v_i}\, e^{-a}}{\sqrt{\dfrac{v_f}{v_i} + e^{-a}}} \tag{25}$$

$$\frac{v_f}{v_i} \infty \frac{I_f\,(1 - e^{-\mu_f \tau})}{I_i\,(1 - e^{-\mu_i \tau})} \tag{26}$$

is which τ is the thickness of the emulsion. The valve of $\dfrac{v_f}{v_i}$ becomes smaller as τ is smaller. For $\tau \to 0, \dfrac{v_f}{v_i} \to \dfrac{\mu_f}{\mu_i}$, but then the sensitivity is zero.

In general the thickness of the film must not be larger than that desirable for recording the soft radiation.

For $\lambda = 10\,\text{Å}$ the Maximum Resolution film absorbs 80 % of the radiation, while the absorption for $\lambda = 2\,\text{Å}$ is only 10 % (Combée and Recourt[2])). Indeed, if we want to use the long wave part of the spectrum $(\lambda \geqslant 10\,\text{Å})$ we must use ultra fine grain films.

§ 5. The filmquality.

With respect to the filmquality (cf. chapter IV), some points will be noted, indicating that the ultra fine grain film may be better than the coarser one.

The coarse grain film is obtained by "ripening" the emulsion. The resulting large grains have grown at the expense of the smaller ones. As this process is statistical, it is clear that σ_Λ (cf. equation (28) chapter IV) increases.

It is not unreasonable to suppose that the quantum yield η of the ultra fine grain film is greater than of the coarser one. The grain diameter of the former amounts to some 500 Å, so that photo-electrons produced in a grain may reach the surrounding grains (c.f. Ehrenberg and White[5])).

§ 6. The 2X method.

Ultra fine grain film had previously been used in Delft for recording X-ray projection images with the so-called 2X method. It is pointed out in a publication[10]) that by taking a primary magnification $M = 2$ the X-ray source can be twice as large as if $M \gg 2$, to obtain the same X-ray image quality. Burger, Combée and van der Tuuk[1]), however, had long before called attention to the appearance of a minimum in the perceptible detail size if the resolution of the fluorescent screen is of the same order of magnitude as the source diameter. Haine and Mulvey[6]) in a publication about recording Fresnel fringes pointed out the possible gain of a 2X magnification, but doubted its realizability.

Unaware of these developments the 2X method was studied in Delft, and led to the conclusion that for soft radiation and a reasonable resolving power the method has some practical advantages. The gain in intensity resulting from the use of a larger source led to exposure times of the order of 10 sec. when using Au white radiation, and to exposure times of about 2 mins with 6 kV and an Al target.

During the Symposium on X-ray Microscopy and Microradiography at Cambridge Cosslett[3]) criticized the 2X method, a criticism based on a misconception of the resolving power of a film, and on an incorrect comparison with the pure projection method. It will be shown that the intensity distribution is exactly the same (on a different scale) for the projection method as for the 2X method[11]).

Consider an X-ray source with an arbitrary intensity distribution $F(p, q)$. A specimen at a distance a has a density distribution $\psi(u, v)$. If the dimensions of the source are small compared to $a + b$ (i.e. $p \ll a + b$ and $q \ll a + b$), the intensity distribution $\varphi(x, y)$ on the film may be written as

$$\varphi(x, y) = c \int\limits_{-\infty}^{+\infty} \int\limits_{-\infty}^{+\infty} \psi(u, v)\, F(p, q)\, \mathrm{d}p\, \mathrm{d}q. \tag{27}$$

From fig. 3 it follows that

$$u = a\,\frac{x - p}{a + b} + p = \frac{ax}{a + b} + \frac{bp}{a + b}. \tag{28}$$

$$v = a\,\frac{y - q}{a + b} + q = \frac{ay}{a + b} + \frac{bq}{a + b}. \tag{29}$$

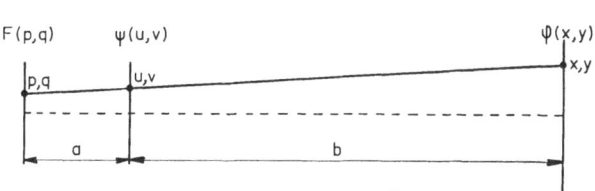

Fig. 3. Coordinate system for source, specimen and film.

87

The new coordinates x_1, y_1 and p_1, q_1 will now be introduced according to

$$x_1 = \frac{a}{a+b}\, x = \eta x, \qquad y_1 = \frac{a}{a+b}\, y = \eta y, \tag{30}$$

$$p_1 = \frac{b}{a+b}\, p = (1-\eta)p, \qquad q_1 = \frac{b}{a+b}\, q = (1-\eta)q. \tag{31}$$

Here

$$\frac{1}{\eta} = \frac{a+b}{a} = M, \tag{32}$$

where M denotes the magnification. Equation (27) then becomes

$$\varphi(Mx_1, My_1) = c \iint_{p_1\, q_1} \psi(p_1 + x_1, q_1 + y_1)\, F\left(\frac{p_1}{1-\eta}, \frac{q_1}{1-\eta}\right) \frac{dp_1\, dq_1}{(1-\eta)^2} \tag{33}$$

The projection method with $\eta \ll 1$ yields the intensity distribution

$$\varphi_p(Mx_1, My_1) = c \int\int \psi(p_1 + x_1, q_1 + y_1)\, F_p(p_1, q_1)\, dp_1\, dq_1. \tag{34}$$

In the same way the 2X method with $\eta = \tfrac{1}{2}$ yields

$$\varphi_{2X}(Mx_1, My_1) = c \iint_{p_1\, q_1} \psi(p_1 + x_1, q_1 + y_1)\, F_{2X}(2p_1, 2q_1)\, 4\, dp_1\, dq_1 \tag{35}$$

Clearly $\varphi_p(Mx_1, My_1) = \varphi_{2X}(Mx_1, My_1)$ for all values of x_1 and y_1 if

$$F_p(p_1, q_1) = F_{2X}(2p_1, 2q_1) \tag{36}$$

i.e. if the intensity distribution of the source in the 2X method is that in the projection method scaled up by a factor 2.

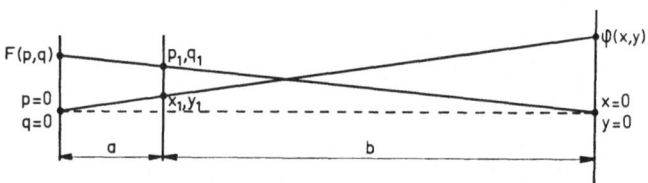

Fig. 4. New variables, introduced for the comparison of the projection method and the 2X method.

Note: The introduction of new variables according to (30) and (31) has a simple physical meaning. Equation (30) means that the X-ray image is projected back into the object plane with $p = 0$ and $q = 0$, the centre of the source, as a projection centre. Equation

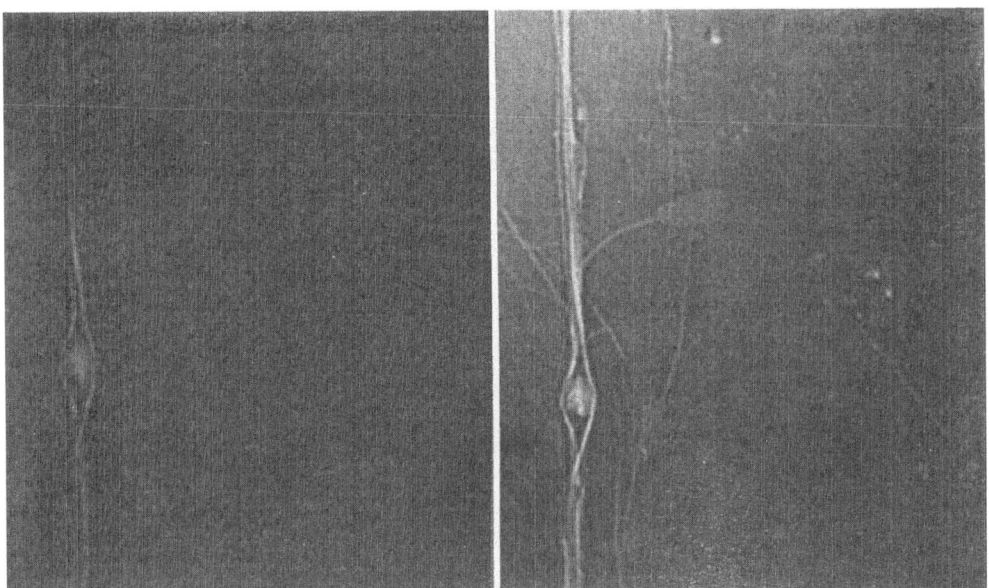

fig. 5
Comparison of Ilford process film (left) with Kodak Maximum resolution film
(right). Bull sperms can only be seen in the right hand picture. Au target, 12 kV,
15 mins exposure, initial magnification 4 X, reproduced at 900 X, negative print.

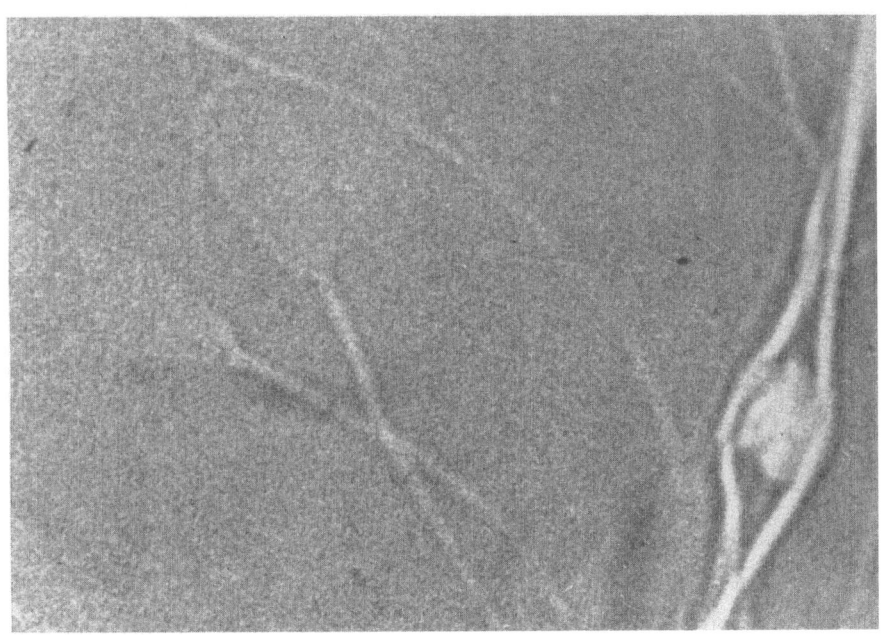

fig. 6
Graininess of Kodak Maximum Resolution film at high optical magnification, (545 X), with the same exposure conditions as in fig. 5.

(31) means that the intensity distribution of the source is projected on the object plane with $x = 0$ and $y = 0$, the centre of the film as a projection centre. The demagnifications are $\eta = a/(a + b)$ and $1 - \eta = b/(a + b)$ respectively (see fig. 4).

From these considerations there follows the general law, valid for contact-, 2X- and projection method: The projection of the image on the object, as seen from the source, does not change if the projection of the source on the object, as seen from the film, remains unchanged.

§ 7. *Further remarks on the 2X method.*

Although the advantages and disadvantages of the 2X method are already discussed in three publications [4, 11, 12]), a further consideration is useful in connection with the new point of view obtained when studying the film properties. In the first place it is shown in (36) that there is a real gain with respect to the projection method as regards intensity and resolving power. With respect to the contact method a gain in resolving power by a factor 2 is doubtful in view of statistics. According to Burger, Combée and van der Tuuk[1]) and Newell[9]) we can suppose this gain to be $\sqrt{2}$. As for intensity the contact method is no doubt superior. The possibility of using very low voltages with reasonable exposure times is a proof of this fact.

The good results we got with the 2X method were in no slight measure due to the use of ultra fine grain film. As a consequence of contrast improvement when using this film, the film has a better resolving power. Contrast is furthermore improved by using low voltages, which is possible due to the gain in intensity of a good 6 times [10]).

§ 8. *Results of exposures with ultra fine grain film.*

If we want to use ultra fine grain film for high resolving power the primary magnification must be $M > 2$. In our provisional experiments we took $M \approx 4$ with an optical magnification of the film of 100 to 200 X. In fig. 5 the difference in recording properties of Kodak Maximum Resolution (ultra fine grain) and Ilford Process Film (normal film) is shown. The anode voltage was 12 kV, the target an Au $(0.2\ \mu)$ layer evaporated on $10\ \mu$ Al foil. The exposure times are almost equal, about 15 mins. During exposure with the

fig. 7
Part of a diatom (arachnoidiscus) recorded with a primary magnification of 2 X.
Reproduced at 1000 X. Au target 12 kV, 12 s. exposure. Au shadow, negative
print.

fig. 8
Human muscle, OsO$_4$ fixation, showing striations. Primary magnification ca 4 X, reproduced at ca 900 X, same exposure conditions as in fig. 5.

Process film the object-film space was washed with coal gas. For comparison an exposure was shown in chapter III of the same object using 6 kV and Au on an Al target. (For these exposures the resolution was better). Furthermore some exposures using ultra fine grain films are shown in fig. 6, 7 and 8.

REFERENCES

1) Burger, G. E. C.,   Phil. Tech. Rev. **8** (1946) 321.
   C. Combée and
   J. H. v. d. Tuuk

2)	Combée, B. and A. Recourt	Phil. Tech. Rev. **19** (1957) 185.
3)	Cosslett, V. E.	X-ray Microscopy and Microradiography, Academic press inc., New York (1957) p. 321.
4)	Cosslett, V. E.	Appl. Sci. Res. **B. 7** (1959) 338.
5)	Ehrenberg, W. and M. White	X-ray Microscopy and Microradiography, Academic press inc., New York (1957) p. 213.
6)	Haine, M. E. and T. Mulvey	Nature, **170** (1952) 202.
7)	Mosley, V. M., D. B. Scott and R. W. G. Wyckoff	Science, **124** (1956) 683.
8)	Mosley, V. M., D. B. Scott and R. W. G. Wyckoff	Biochim. Biophys. Acta, **24** (1957) 235.
9)	Newell, R. R.	Brit. Journ. Rad. **30** (1938) 493.
10)	Ong Sing Poen and J. B. Le Poole	Appl. Sci. Res. **B, V** (1956) 543.
11)	Ong Sing Poen and J. B. Le Poole	Appl. Sci. Res. **B. 7** (1958) 265.
12)	Ong Sing Poen and J. B. Le Poole	Appl. Sci. Res. **B 7** (1958) 343.

CHAPTER VI

PRACTICAL X-RAY MICROSCOPY

§ 1. *Introduction.*

The information we obtain with the aid of an X-ray microscope mainly concerns the structure of the object. The contrast obtained results from differences in mass per unit area ϱz, and chemical composition, expressed in the absorption coefficient μ_λ. The determination of these quantities is called mass analysis, and chemical analysis respectively. Chemical analysis can be either quantitative or qualitative. As the eye is not suited for measuring intensities of radiation the quantitative analysis must be carried out with objective instruments. This necessitates image dissection, which may be followed by a synthesis (e.g. scanning microscope). The possible methods of arriving at such analysis have been investigated among others by Engström and Lindström [6]; Zeits and Baez [2]; Long and Cosslett[4]; Wallgren[12]; Wyckoff and Mosley[9].

In Delft the development of practical X-ray microscopy is aimed mainly at information about the structure of the object without further analysis of the image. This has several causes. In the first place some industries (especially the paper industry) were interested in this particular way of carrying out X-ray microscopy; secondly the development of the microscope itself was considered the most important task. As a third cause might be mentioned the lack of co-operators interested in the problems of quantitative microscopy. According to Isings [7, 8] projection X-ray microscopy is in many cases the indicated method to reach a solution of the problems emerging from paper research. The composition of this material (fibres, fillers and dirt) is sufficiently known, or at least easily determined by other methods. One of the things one wants to know is the spatial distribution of the various components. Furthermore some idea of the kind and extent of damage to the fibres and their mutual attachment is desired. If the research is carried out with an optical microscope refraction, scattering and absorption will blur the essential details while the small depth of focus is a great handicap. Admittedly cutting thin sections could reduce the effects just mentioned, but the spatial structure is lost in this process. With the large penetrating power of the radiation, the large depth of focus and the practical absence of scattering and refraction, the projection X-ray microscope is pre-eminently suitable for this kind of research.

This chapter will therefore deal principally with the possibilities and practical applications connected with the large depth of focus and penetrating power of the radiation.

§ 2.   *The adaptation of the wavelength to the object.*

If mechanical damage to a given object with an average thickness z and average absorption coefficient μ_λ is not permitted we may question whether a wavelength λ exists, giving maximum information; in other words, does a suitable choice of wavelength for a given object show an optimum in detail in the image? To answer this question we consider figure 1 a.

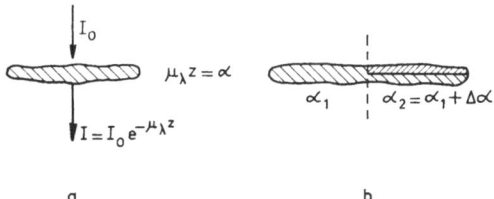

fig. 1
Computation of the optimum absorption exponent a.

The transmitted intensity can be written as

$$I = I_o \, e^{-\mu_\lambda z} \tag{1}$$

in which I_o is the intensity of the incident radiation. In the regions between absorption edges the variation of μ_λ with wavelength can be written as

$$\mu_\lambda = c \, \lambda^k \tag{2}$$

in which c is a constant for a given region of wavelength and k is also a constant which is roughly 3. The wanted information (see

§ 1 of this chapter) is connected with a variation in the product

$$\mu_\lambda z = a \tag{3}$$

In the first instance we are not interested in whether a variation Δa is a result of a change in μ_λ or of a change in z. The problem is to choose λ such that at a given specimen the smallest variation Δa can be determined. Consider two regions with absorption exponents respectively a_1 and a_2 (see fig. 1 b), in which

$$a_2 = a_1 + \Delta a \tag{4}$$

According to (2) and (3) we may write

$$a_1 = c_1 \lambda^k z_1 \tag{5}$$

and

$$a_2 = c_2 \lambda^k z_2 \tag{6}$$

Hence

$$\Delta a = \lambda^k (c_2 z_2 - c_1 z_1) \tag{7}$$

and

$$\frac{\Delta a}{a} = \frac{c_2 z_2 - c_1 z_1}{c z} \tag{8}$$

from which we see that $\dfrac{\Delta a}{a}$ is independent of λ. This means that this magnitude is a natural magnitude of the object itself which therefore determines the detail in question. This is not the case for the magnitude Δa as according to (3)

$$\Delta a = \mu_\lambda \Delta z + z \Delta \mu_\lambda \tag{9}$$

in which μ_λ depends on λ also.

Note: The values of c, c_1 and c_2 are different for $\lambda < \lambda_e$ and $\lambda > \lambda_e$ in which λ_e is the wavelength of an absorption edge.

In view of the above we shall call $\dfrac{\Delta a}{a}$ the object detail. To determine this we can make use of long wavelength or short wavelength radiation. Although $\dfrac{\Delta a}{a}$ is independent of λ it is possible that its value can be obtained in a shorter time by a suitable choice of λ. Or, if we have a certain number of X-ray quanta at our disposal, it is possible that by a correct choice of the quantum energy a minimum value of $\dfrac{\Delta a}{a}$ can be determined. Instead of determining the most suitable wavelength we can determine the optimum value of α, α_{opt}. The advantage is that from the calculated value of α_{opt} the corresponding wavelength λ_{opt} can be determined easily by measuring the absorption. Assuming that ξ_0 quanta impinge on an object element, the number of quanta transmitted per element amounts to

$$\xi = \xi_0 e^{-\alpha} \tag{10}$$

or

$$\alpha = -\ln \xi + \ln \xi_0 \tag{11}$$

and

$$\Delta \alpha = -\frac{\Delta \xi}{\xi} \tag{12}$$

As a consequence of statistical fluctuation in the number of quanta and the film (see chapter IV, § 4) the following equation must be satisfied, viz.

$$\Delta \xi = \pm q \sqrt{\xi} \tag{13}$$

for a variation in $\Delta \xi$ to be considered with reasonable certainty to be a consequence of $\Delta \alpha$. Here q is a constant which also depends on the quality of the recording medium.

Equations (12) and (13) together give the condition

$$\Delta \alpha = \frac{q}{\sqrt{\xi}} \tag{14}$$

Substitution of (10) gives

$$\Delta \alpha = \frac{q \, e^{1/2 \alpha}}{\sqrt{\xi_0}} \tag{15}$$

For a given value of ξ_0 the object detail is

$$\frac{\Delta \alpha}{\alpha} = \frac{q \, e^{1/2 \alpha}}{\alpha \sqrt{\xi_0}} \tag{16}$$

This has a minimum for $\dfrac{d}{d\alpha} \left(\dfrac{\Delta \alpha}{\alpha} \right) = 0$

or

$$\frac{1/2 \, \alpha \, e^{1/2 \alpha} - e^{1/2 \alpha}}{\alpha^2} = 0$$

Hence

$$\alpha_{opt} = 2 \tag{17}$$

For a certain value of $\dfrac{\varDelta a}{a}$ the necessary number of incident quanta is according to (16)

$$\xi_{\,o} = \frac{q^2\,e^a}{\left(\dfrac{\varDelta a}{a}\right)^2 a^2} \tag{18}$$

The function (16) is represented in fig. 2. In the same diagram the number of incident X-ray quanta per unit area X_o is drawn.

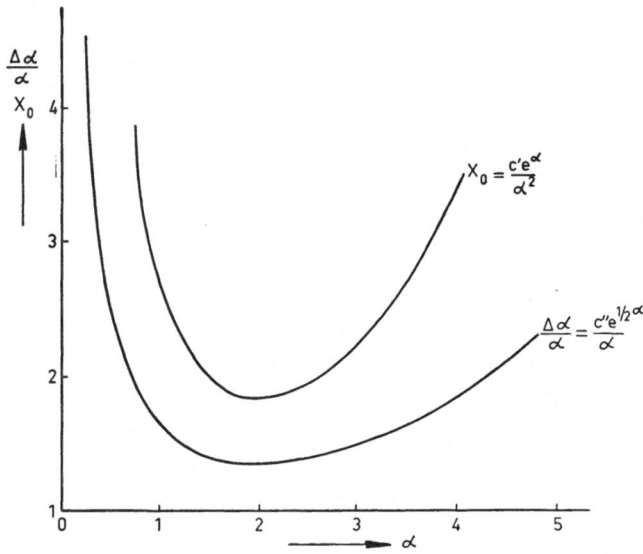

fig. 2

Number of X-ray quanta per unit area X_0 and the object detail $\varDelta a/a$ plotted against a.

For the projection microscope however, it is not the value of ξ_o that is particularly important, but the exposure time t. The relation between these is

$$\xi_{\,o} \infty \frac{I_x\,t}{h\nu} \infty I_x\,t\,\lambda \tag{19}$$

in which I_x is the total X-ray output and $h\nu \infty h/\lambda$ the energy of one quantum. According to chapter II § 9 equation (23),

$$I_x \infty V^3 \tag{20}$$

in which V is the anode voltage. The "wavelength" of the white radiation, i.e. the maximum of the spectral intensity distribution, is inversely proportional to the voltage, so that for (19) we may write

$$\xi_o \propto \frac{t}{\lambda^2}$$

With (2) and (3) for $k = 3$ this gives

$$\xi_o \propto \frac{t}{a^{2/3}} \tag{21}$$

Substitution of which in (16) gives

$$\frac{\Delta a}{a} \propto \frac{e^{1/2 a}}{a^{2/3} t^{1/2}} \tag{22}$$

This has a minimum for $a = {}^4/_3$, corresponding to an absorption of 73,4 %.

For some (e.g. living) objects, it is important to keep the X-ray dose as small as possible. For the dose we can write

$$\chi \propto \frac{\xi_o (1 - e^{-a})}{\lambda} \tag{23}$$

with $k = 3$ (eq. 2) this becomes

$$\chi \propto \frac{\xi_o (1 - e^{-a})}{a^{1/3}} \tag{24}$$

The value of ξ_o can be derived from eq. 15, so

$$\xi_o = \frac{p^2 e^a}{(\Delta a)^2}. \tag{25}$$

Hence

$$\chi \propto \frac{e^a (1 - e^{-a})}{(\Delta a)^2 a^{1/3}} \tag{26}$$

or

$$\chi \propto \left(\frac{a}{\Delta a}\right)^2 \frac{e^a (1 - e^{-a})}{a^{7/3}} = \left(\frac{a}{a \Delta}\right)^2 \frac{(e^a - 1)}{a^{7/3}} \tag{27}$$

For a certain object detail $\frac{a}{\Delta a}$, χ will show a minimum if

$$a^{7/3} e^a - {}^7/_3 a^{4/3} (e^a - 1) = 0$$

or

$$a e^a - {}^7/_3 e^a + 1 = 0 \tag{28}$$

So

$$a_{opt} \approx 2.2,$$

corresponding to an absorption of some 90 %.

§ 3. *Determination of the thickness of a section.*

Suppose that the thickness z and the average absorption coefficient μ^λ of an object are given. In the object there are parts with an absorption coefficient $\mu_\lambda + \Delta\mu_\lambda$. To make the parts visible we can either observe the complete object or cut it into sections and observe them separately. Leaving out of consideration the time needed for cutting and preparing, we want to investigate for which method the total exposure time is shorter.

Suppose the object is cut into n equal sections (see fig. 3).

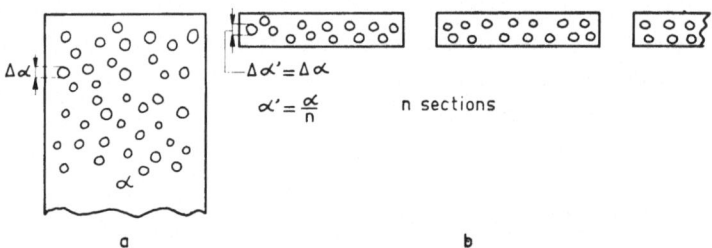

$$-\Delta\alpha' = \Delta\alpha$$
$$\alpha' = \frac{\alpha}{n} \qquad n \text{ sections}$$

a   b

fig. 3
Determination of the section thickness.

For the complete specimen, the exposure time required to make the details visible (according to equation (22)) amounts to

$$t_1 \propto \left(\frac{a}{\Delta a}\right)^2 \frac{e^a}{a^{4/3}} \tag{29}$$

If the sections, for every value of n, are observed under optimum conditions we may write for the exposure time of one section (see fig. 3)

$$t_n \propto \left(\frac{a'}{\Delta a'}\right)^2 \propto \left(\frac{a}{\Delta a}\right)^2 \frac{1}{n^2} \tag{30}$$

As has already been shown in § 2, $\dfrac{\Delta a}{a}$ is a constant. The total exposure time for observing n sections then amounts to

$$t_{tot} = n\, t_n = \frac{t_1}{n} \tag{31}$$

100

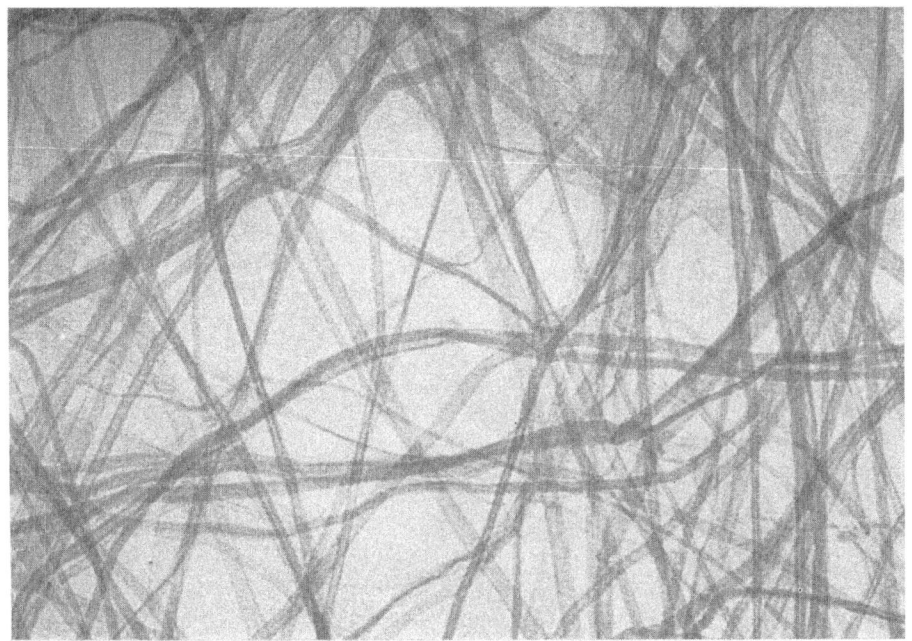

fig. 4
Tissue paper, untreated. Au target 12 kV, magn. appr. 125 X.

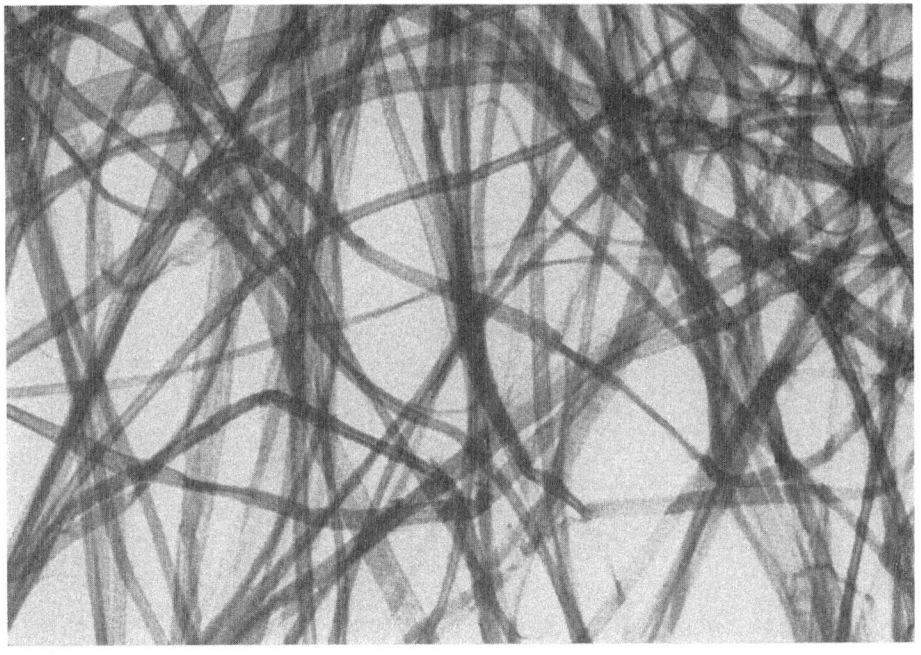

fig. 5
Tissue paper, treated with an alcoholic Iodine solution. Au target 12 kV, magn.
appr. 125 X.

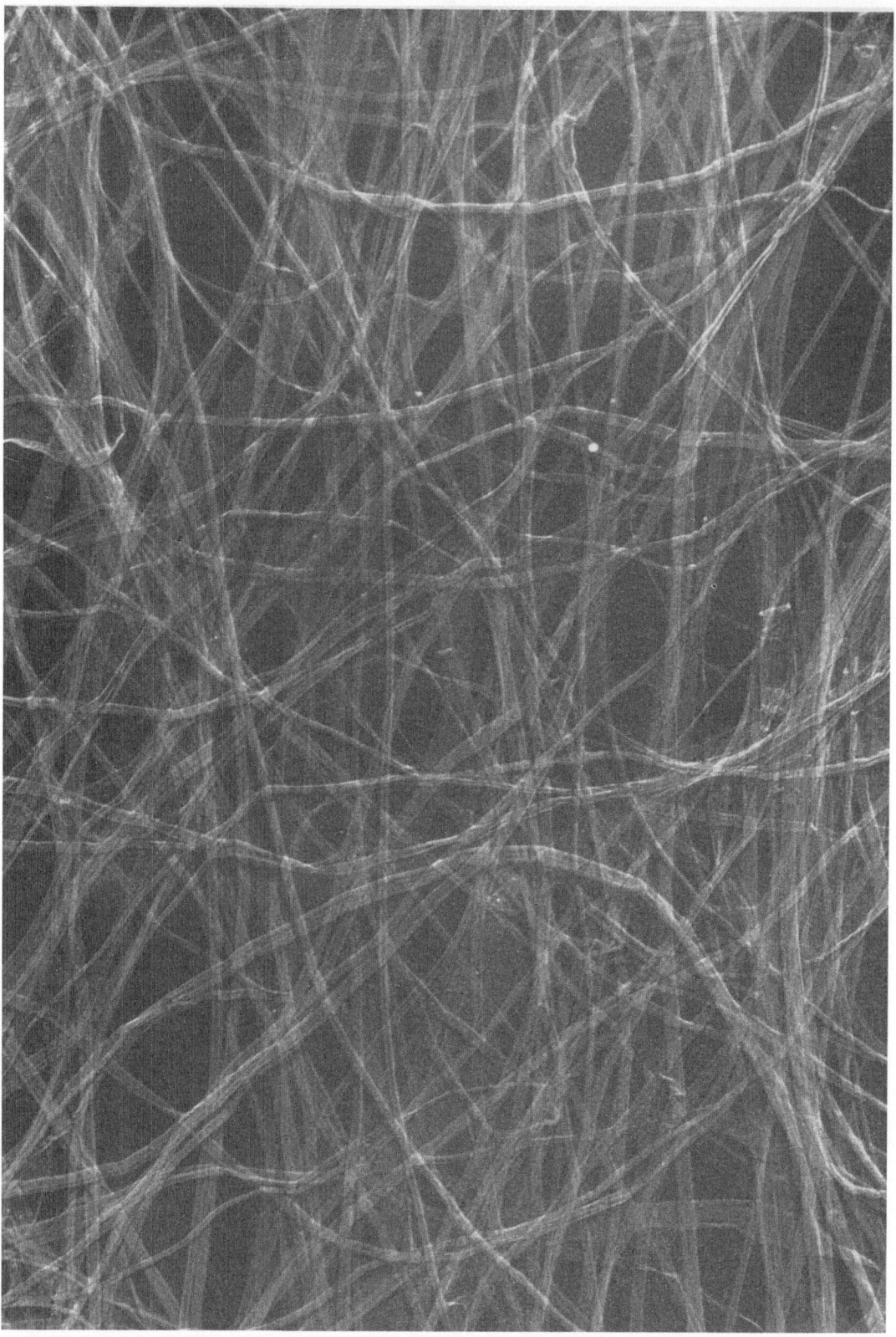

fig. 6
Tissue paper, gold shadowed. Membranous structure is distinguishable. Au target,

From equation (31) follows the preferability of cutting the specimen in a large number of sections and observing them separately with the adapted wavelength. In general the use of soft radiation is necessary in this case. If the choice of wavelength is limited, the thickness of the section must be such that $a = \frac{4}{3}$. Although we have derived equation (25) for projection microscopy in particular, this relation holds generally, provided a equals the optimum value.

§ 4. *Preparation methods.*

In most cases the choice of the maximum effective wavelength is severely limited. The value of a can be increased, however, by an appropriate treatment of the object. Some examples of preparation technique applied in Delft with favourable results will be described briefly here.

a) Direct staining of the object.

To make the fibres in paper clearly visible the object can be treated with an alcoholic iodine solution. The degree of staining can be controlled by the iodine concentration of the solution. Some kinds of fibre can be stained better with J-KJ solution. (Isings [8])). In fig. 4 and 5 contrast improvement by means of iodine staining is quite well visible. The object is lens tissue.

b) Shadowcasting with a heavy element.

This technique, often applied in electron microscopy, can be used here when we are interested in the structure of the surface. The shadow layer must be thicker of course than that used in electron microscopy. Shadowcasting gives information which in general is not specific to X-ray microscopy. So application of this technique is limited to cases for which a certain characteristic of the projection method is desired, e.g. for making stereoscopic exposures at high magnifications. An example of shadowcasting can be seen in fig. 6.

c) Selective staining.

By treating the object with certain agents containing heavy elements selective staining may occur. An example of this method is treating skin sections with OsO_4. This fixing agent often used for biological specimens is reduced especially by unsaturated fatty acids. After washing, the Os remains. Fig. 7 shows a

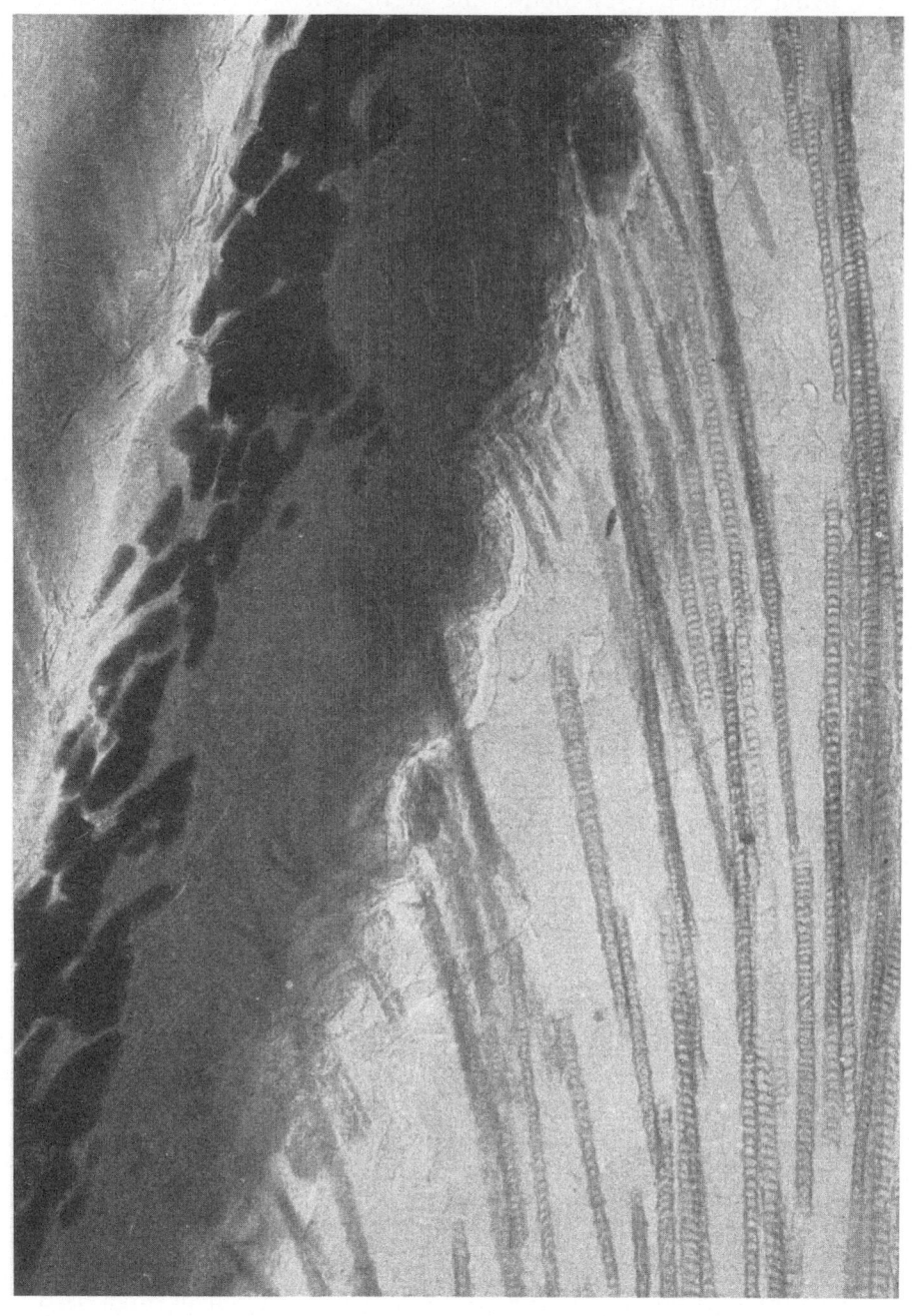

fig. 7
A section of mouse skin, stained with OsO_4. Fat particles can clearly be seen.

fig. 8
Negative replica of human skin, gold shadowed. Au target, 12 kV, magn. appr. 85 X, negative print.

picture of such a skin section. The technique of selective staining for X-ray microscopy is still in its initial stage. Much can be attained in this field, especially in combination with the use of monochromatic radiations of various wavelengths. For example the object can be treated with different staining liquids, each of them staining different parts. By making several exposures with different wavelengths the different elements can be shown. The photographs can be projected superimposed after having been suitably coloured. The use of colours to mark elements has been demonstrated by Cosslett and Duncumb[5]). They used images originating from a scanning microscope.

d) Replication.
Replicas are made to observe coarse surfaces of thick, strongly absorbing specimens. This technique too is known from electron microscopy. For X-ray microscopy the replicas of course may

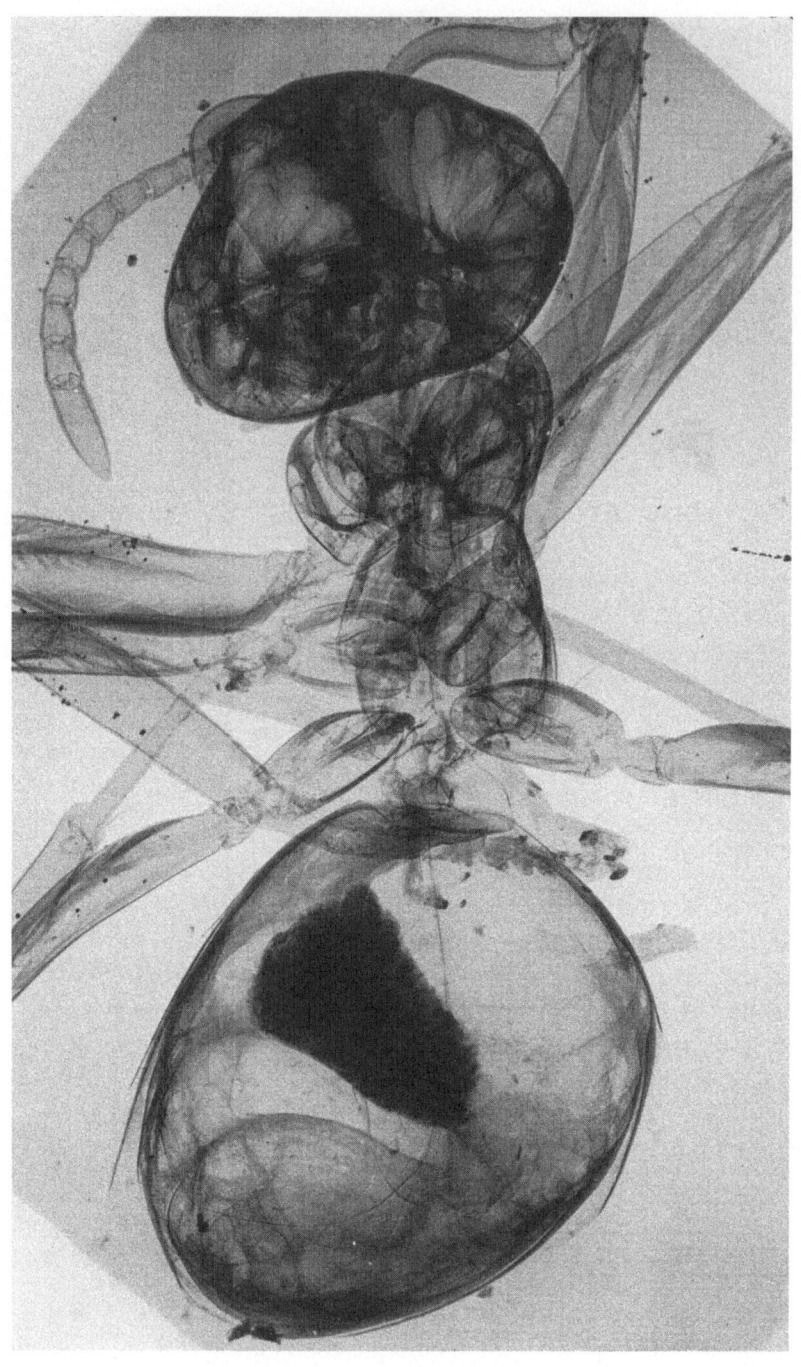

fig. 9
An ant, after a sugar-leadacetate meal. Au target. 12 kV, magn. appr. 60 X.

be rather thick. The "image" originates mainly from the shadowcasting (see under b). Studying replicas with the aid of a projection X-ray microscope has advantages over observing the object itself. Besides stereoscopy being possible at large magnifications the image is more readily surveyed and inconvenient reflections and refraction are eliminated. When studying fracture surfaces etc. this method can render good service[10]). An example is shown in fig. 8.

e) Contrast agents.

Administering contrast agents, a technique often applied in medical X-ray diagnostics can also be applied in projection microscopy. De Groot[3]) for instance made contact exposures of bees which had consumed sugared water containing $BaSO_4$. A powerful agent, leadacelate-sugar solution, was applied to an ant, the result being given in fig. 9.

§ 5. *Stereo-microscopy.*

In microstereography the aim is to see the object in the right spatial proportions, but on an M times larger scale. If the object has the spatial coordinates x, y and z we want to get the impression, when observing the stereo photograph, that the specimen has the spatial coordinates Mx, My and Mz, in which M is the magnification. The geometrical conditions that must be satisfied can be deduced easily from fig. 10 (See also Reynders[11]) and Albada[1])).

In fig. 10 a the M times magnified object is placed at a distance l from the observer.
The eyes subtend an angle φ at the object, where $\tan \frac{1}{2}\varphi \approx e/2\,l$. Here e is the inter-ocular distance which is normally 65 mm. In fig. 10 b the object can be thought of as replaced by two film images obtained by projecting the various points of the object with the eye pupils as projection centres. Fig. 10 e shows that the same image on the film can also be the result of two equal objects which are M times smaler with coordinates x, y and z. As the two objects are exactly identical, the two images can also be thought of as projections of one object, the projection centre for which is moved over a distance Δs (fig. 10 d). Fig. 10 e shows that instead of moving the projection centre the object can be moved over a distance $\Delta a = \Delta s$.

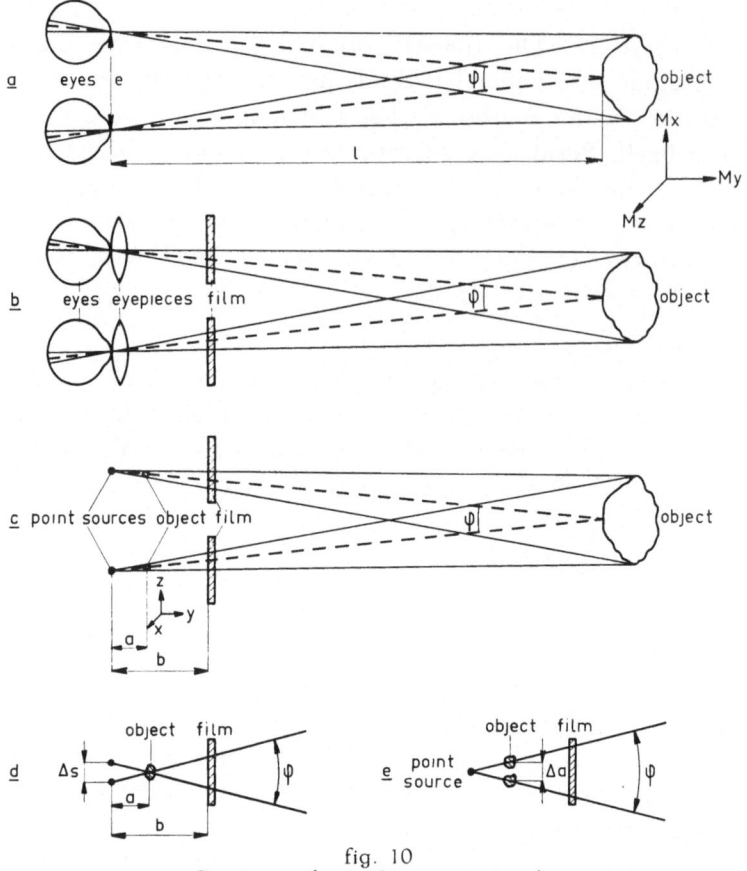

fig. 10
Conditions for making stereographs.

Furthermore it can be easily recognized that for these movements
the relationship

$$\varDelta s = \varDelta a = \frac{ac}{l} \tag{32}$$

must hold. In general the distance a is not known exactly, especially for large magnifications for which the object is close to the
source. The value of b (see fig. 10 c), on the other hand, can be
measured accurately and is often a constant of the apparatus. By
introducing the film magnification

$$M_f = \frac{b}{a} \tag{33}$$

we can write for (32)

$$\varDelta s = \varDelta a = \frac{be}{M_f \, l} \tag{34}$$

In which $e = 65$ mm and $l \geqslant 250$ mm. For thick objects it is desirable to make $l > 250$ mm. The image becomes more readily surveyable because the eye axes when observing the various points need only turn through small angles with respect to one another.

If we write (34) as

$$M_f \, \Delta s = M_f \, \Delta a = \frac{be}{l} \tag{35}$$

the second member is a constant. The antecedent is the product of magnification on screen or film and displacement. To satisfy condition (32) the object or the source must be moved in such a way that the image is displaced over a distance $\frac{be}{l}$ Therefore it is not necessary to know the displacement Δa or the absolute value of a. If we are able to measure Δa, e.g. with a calibrated specimen movement, then with (35) the magnification M_f can be determined (§ 6).

Besides translating the object or the source, the object can also be tilted (fig. 11) which has certain advantages over shifting.

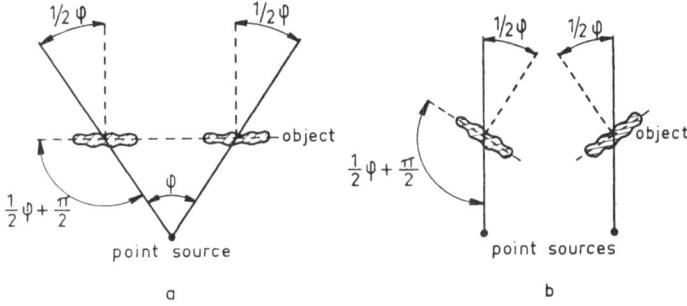

fig. 11
The conditions for translating (a) and tilting (b) the specimen.

In this case there is less loss of field of view as appears below (fig. 12). Fig. 12 a represents the case of the object being translated. The double hatched part of the object occurs in both stereo photographs and can therefore be viewed stereoscopically. The loss of angular field, amounts to

$$2 \, \Delta \gamma_t = 2 \varphi \tag{36}$$

109

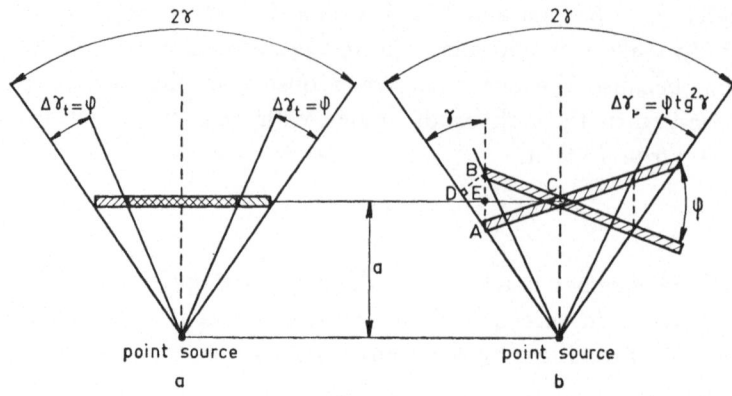

fig. 12.

Comparison of the field of view in the case of translating (a) and tilting (b) the specimen.

From. fig. 12b it follows thas $EC \approx a \tan \gamma$; $BA \approx a \varphi \tan \gamma$, and $DB \approx a\varphi \tan\gamma \sin\gamma$; $DS = \dfrac{a}{\cos \gamma}$ so that $\varDelta\gamma \approx \dfrac{DB}{DS} \approx \dfrac{DB}{a} \cos \gamma$, so that the loss of field amounts to

$$2\varDelta\gamma_r = 2\varphi \sin^2 \gamma \qquad (37)$$

In general $\sin^2 \gamma << 1$. In the case of $\gamma = 15°$ $\sin^2 \gamma = 0{,}066$. A disadvantage of tilting is that the axis of tilt cannot be accurately determined so that the result is a combination of tilting and translating. Furthermore tilting can only be carried out correctly if the source to object distance is not too small. A special advantage of translating is the possibility of obtaining a physically justified measurement of the magnification.

§ 6. *Determination of the magnification.*

As was mentioned briefly in chapter II we can hardly speak of "a magnification" of thick objects as its value is different for the different points of the object. For thin, plane objects the magnification can be determined in several ways.

a) By taking a picture of a reference object in the same photograph, or by comparison with the dimensions of the meshes of the specimen carrier. A fine silver grid (1500 mesh/inch) is often used for this purpose. The same grids as for the electron microscope can be used as specimen carriers, since the dimensions of these are well defined.

110

b) By measuring the object with the aid of an optical microscope. If the irradiated part cannot be localized it is very difficult to find the detail in question in a light microscope, especially because the complete image often looks quite different.

c) From the known displacement of the object with stereo-exposures. If stereo-exposures have been made, a magnification standard can also be given for thick objects. The accuracy of the magnification is the same as that of the displacement. At an object displacement of Δa the various image points are displaced over a distance $M_f \Delta a$ in which M_f is the magnification (equation 35). For a known value of Δa the magnification M_f can be determined from $M_f \Delta a$, which can be done by superimposing the negatives with the boundaries coinciding.

As this method for measuring the magnification is the only generally reliable way, all projection microscopes should be provided with means to obtain a calibrated object translation.

REFERENCES

1) Albada, L. E. W. v.   Handbuch der wissenschaftlichen und angewandten Photographie, Verlag von Julius Springer, Wien (1931) Band VI.
2) Baez, A. V. and Zeits, L.   X-ray microscopy and microradiography, Academic press inc. New York (1957) p. 417.
3) Botden, P. J. M., B. Combée and J. Houtman   Phil. Tech. Rev. 14 (1952) 114.
4) Cosslett, V. E. and J. V. P. Long   X-ray microscopy and microradiography, Academic press inc., New York (1957) p. 435.
5) Cosslett, V. E. and Duncumb   X-ray microscopy and microradiography, Academic press inc., New York (1957) p. 374.
6) Engström, A. and B. Lindström   Experientia, 3 (1947) 191.
7) Isings, J., G. van Nederveen, Ong Sing Poen and J. B. Le Poole   Proc. Stockholm Conf. El. Micr., Almqvist Wiksell, Stockholm (1956) p. 282.
8) Isings, J.   (Centraal Laboratorium T.N.O., Delft) personal communication.
9) Mosley, V. M., and W. G. Wyckoff   Jour. Ultrastructure Res. 1 (1958) 337.
10) Ong Sing Poen and J. B. Le Poole   X-ray microscopy and microradiography, Academic press inc., New York (1957) p. 91.
11) Reynders, F. H.   Thesis, Utrecht, 1951.
12) Wallgren, G.   X-ray microscopy and microradiography, Academic press inc., New York (1957) p. 461.

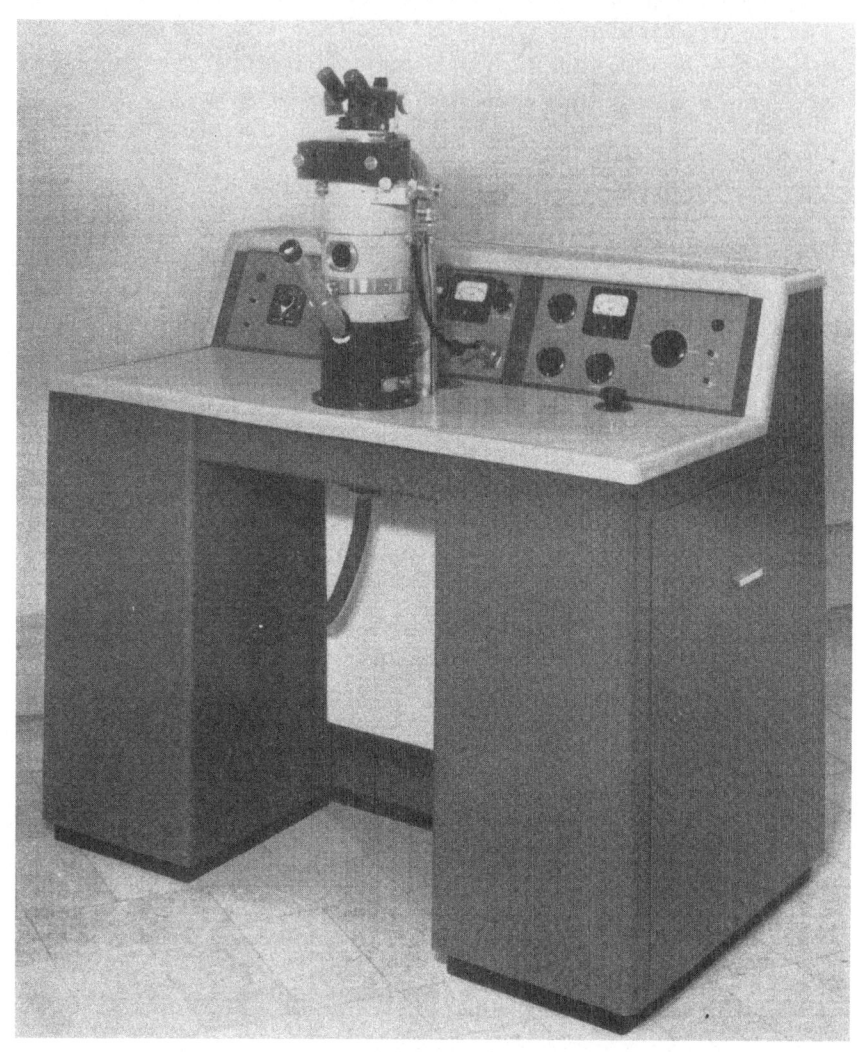

fig. 1.
Commercial projection X-ray microscope, manufactured by the Electron Microscope Division of the Technical Physics Department T.N.O. and T.H.

CHAPTER VII

A COMMERCIAL PROJECTION MICROSCOPE

§ 1. *Introduction.*

Although the development of the projection microscope in Delft was mainly aimed at the improvement of the apparatus itself, gradually the need for a reliable microscope which would be continuously available for studying the application possibilities was felt. Early in 1958 the design of such a microscope was started. When in November of that year the State University of Ghent gave an order for the building of a similar apparatus, it was decided to sell the first microscope. This apparatus, of which in fig. 1 a photograph and in fig. 2 a schematic plan are given, will be described briefly in this chapter.

§ 2. *The voltage range.*

The design of this apparatus is aimed at the largest possible voltage range without a disproportionate increase of cost. The lowest voltage is determined by the admissible exposure times. On account of our experiences with the experimental apparatus we took as lowest usable voltage about 5 kV. The choice of the highest voltage is determined by the cost of high tension and lens current installations. Above about 20 kV the cost increases more than in accordance with the increase of the application possibilities, and therefore this value is taken as maximum. The anode voltage is adjustable in 4 steps. When switching over to another voltage the lens currents are automatically adjusted to such values that the lens powers stay nearly unchanged.

§ 3. *The object space.*

When using high voltages it is necessary to make the target very thin to obtain a good resolving power. For 10 kV the depth of

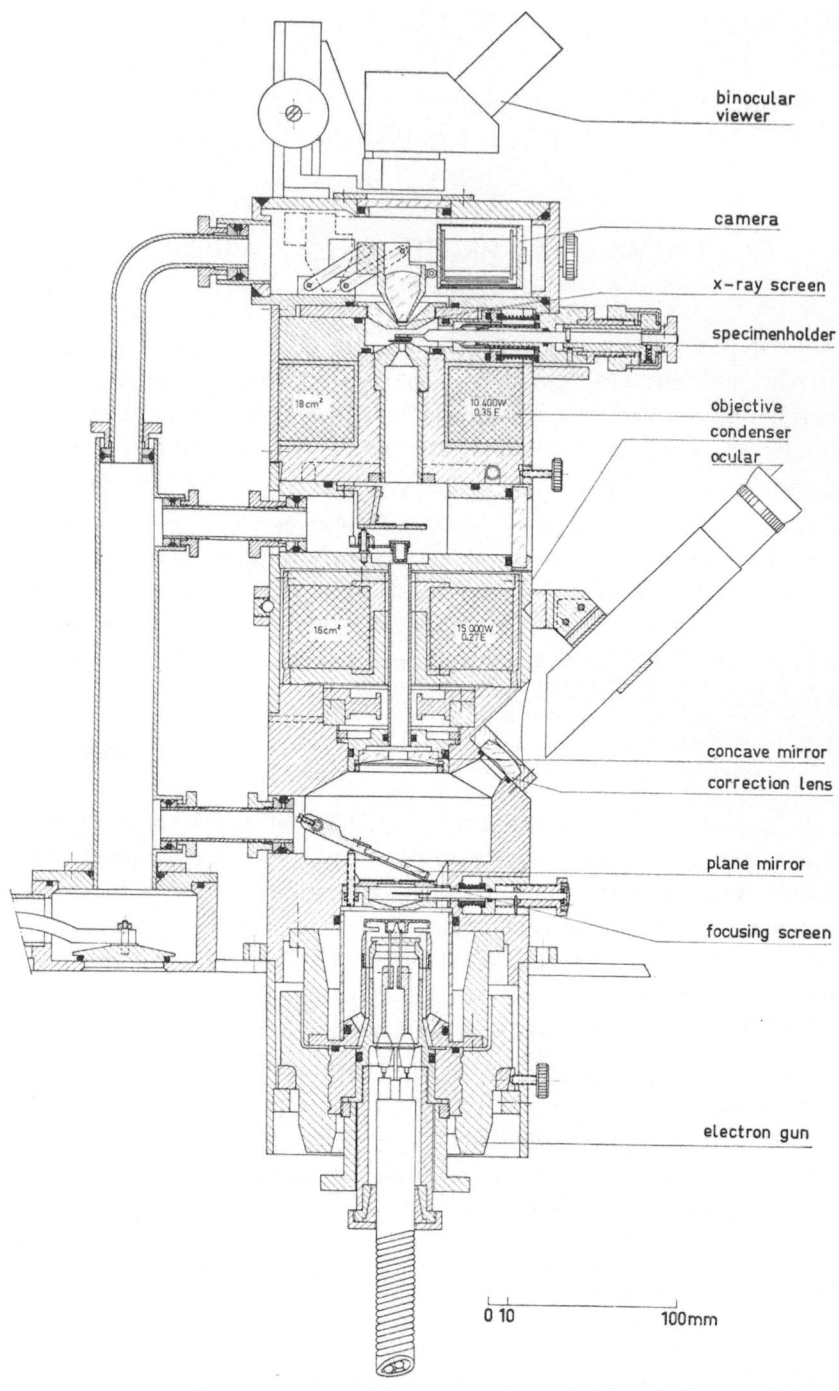

binocular viewer

camera

x-ray screen

specimenholder

objective
condenser
ocular

concave mirror

correction lens

plane mirror

focusing screen

electron gun

18 cm²

10 400 W
0.35 E

16 cm²

15 000 W
0.27 E

0 10   100mm

fig. 2. Simplified cross-section of the commercial projection microscope.

penetration of the electrons in Au according to the Thomson-Whiddington law [4]) amounts to about 1 μ. The use of a target which is thinner than 1 μ presents serious difficulties if it functions also as vacuum seal. As furthermore when using low voltages the absorption of air plays a part we decided to place object and film in vacuo. The fact that only dry objects can be studied is not considered to be a serious disadvantage.

§ 4. *Interchangeable target.*

Although up till now the possibilities of chemical analyses have not been utilized in Delft, we are convinced of their importance. For this kind of work the microscope must provide monochromatic rays of various wavelengths, which can be obtained by using different materials as targets. In our microscope we can make a choice of 4 different materials, the exchange of which can take place without breaking the vacuum. Small shifts in a plane perpendicular to the optical axis make the selection of a suitable part of the target possible. As shown in chapter III this is necessary in connection with roughening and crater formation.

§ 5. *Focusing and centring.*

The focusing method as described in chapter III is also applied here. Centring of the electron optical system takes place with the aid of the secondary image. In contrast to the experimental apparatus, here the secondary image is viewed with mirror optics. Due to the rather large aperture (0.3) some correction is necessary. The corrective lens (see Hekker[1])) functions at the same time as vacuum seal. The objective magnification amounts to 5 X, the total magnification to about 25 X.

§ 6. *The X-ray fluorescent screen.*

As focusing does not take place with the aid of the fluorescent image and as the apparatus is unsuitable for visual observation, the fluorescent image plays a secondary part. It is necessary however to have an indication of whether the object is in the image field. The fluorescent screen combined with the viewer should be considered as a view-finder. Here the magnification is not so important as the image brightness, which is raised at the expense of the former. The fluorescent material is put on an aplanat. The total magnification of this finder system is about 20 X. A simple arrangement makes the screen move away when the camera is in the exposure position. Another arrangement for moving the screen away independently of the position of the camera will be made, so that the object itself can be studied with the viewer.

§ 7.   *The camera.*

The camera is designed for 20 exposures on 35 mm film. Film transport takes place automatically during the shifting of the camera. A shutter is opened when the camera is moved into the exposure position, which would, however, produce complications for very short exposure times. In this case the electron beam must be interrupted, which can be done with the aid of the intermediate target. The magnification on the film is about 4 X larger than on the screen; the image field has a diameter of about 25 mm ($2\,\alpha \approx 36°$).

§ 8.   *Specimen holder.*

The specimen holder is movable in 3 directions, and can be moved over small distances by a calibrated shift for making stereo

fig. 3
Close up of the target-, aperture- and specimenholder.

116

exposures (see fig. 3). From the known displacement the magnification can be determined, as has already been discussed in chapter V/I. The magnification on the film is adjustable between 150 and 10 X; the admissible optical magnification of the film is about 10 X. For realization of the two extreme values of the magnification the specimen must be put very close to the target in the first case, and in the second almost at the upper pole piece. The magnification cannot be varied continuously over the complete range because the specimen holder has a certain thickness. Without breaking the vacuum the specimen bar can be turned through 180° after a small outward shifting. For magnifications of more than 150 X the object must be fixed on the target. The target holder, however, cannot be removed without breaking the vacuum. Another possibility for obtaining large magnifications is to fix the target on the specimen holder, for which purpose the target holder can be shifted back completely. A thin target can be put in to make an observation of the astigmatism with the "forward scattering focusing aid", according to Nixon[3]), possible.

§ 9. The electron lenses.

As targets, diaphragm and object are inserted sideways, the pole piece distance of the objective is made rather large. For a symmetrical lens it would amount to 6 mm whereas the bore amounts to 3 mm. The spherical aberration constant according to the data of Liebmann and Grad[2]) is 1,5 mm, that is about 3 X larger than with our experimental apparatus. While preserving the same spherical aberration the pole piece distance can be enlarged by making the lens asymmetrical. In our microscope this distance is 7 mm whereas the bore of the upper pole piece has a diameter of 4 mm. The objective, which has a focal length of 1,8 mm, consumes about 90 watt, and is water-cooled. The condenser has such dimensions that with an image of the electron source on the intermediate target the image rotation amounts to about 90°, the advantage of which was discussed in chapter III.

The apparatus as described here is being designed and made in close co-operation with D. R. van den Bos of the Technological University, D. D. Groenheyde, G. M. van Koppen, J. Kramer and J. Postma of the Technical Physics Department T.N.O. and T.H. at Delft.

REFERENCES

1) Hekker, F.   On concentric optical systems, Thesis, Delft 1947.
2) Liebmann, G. and   Proc. Phys. Soc. **B 64** (1951) 956.
   E. M. Grad
3) Nixon, W. C.   Int. Conf. on El. Micr. Berlin 1958.
4) Whiddington, R.   Proc. Roy. Soc. **A 89** (1914) 554.

One picture tells more than a thousand words

Chinese proverb

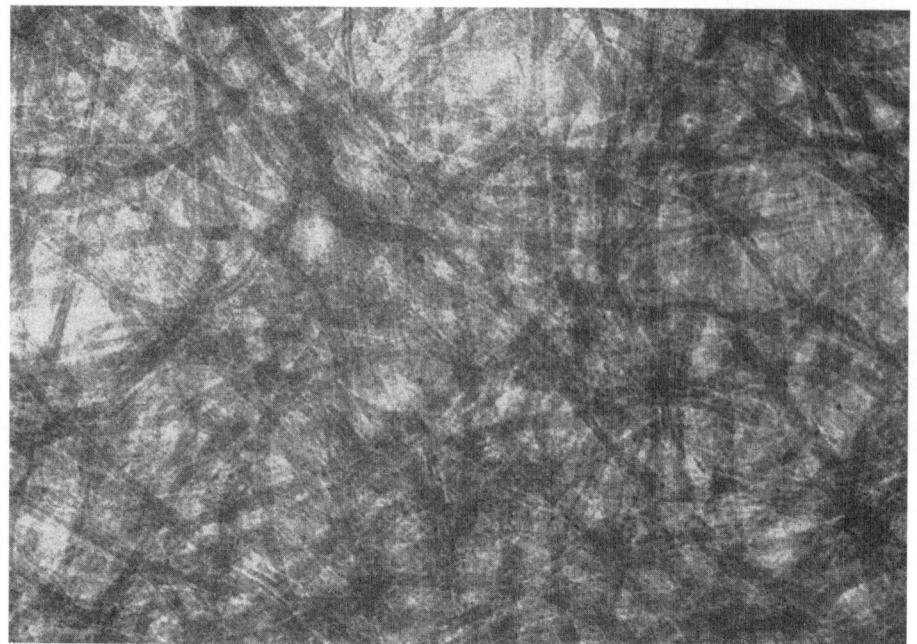

fig. 1. Ordinary paper, magn. ca 130 X

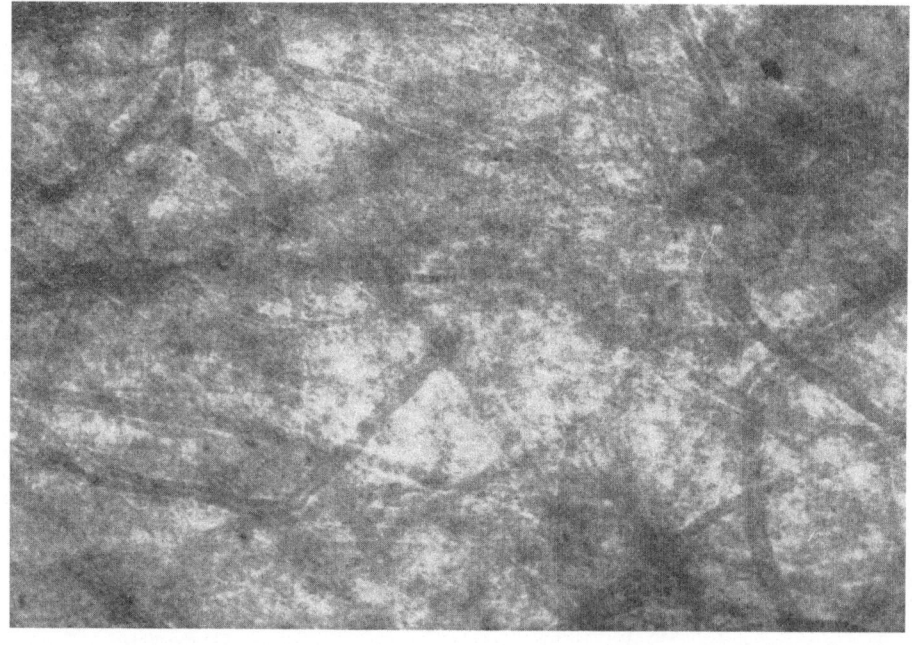

fig. 2. Bible paper, magn. ca 130 X

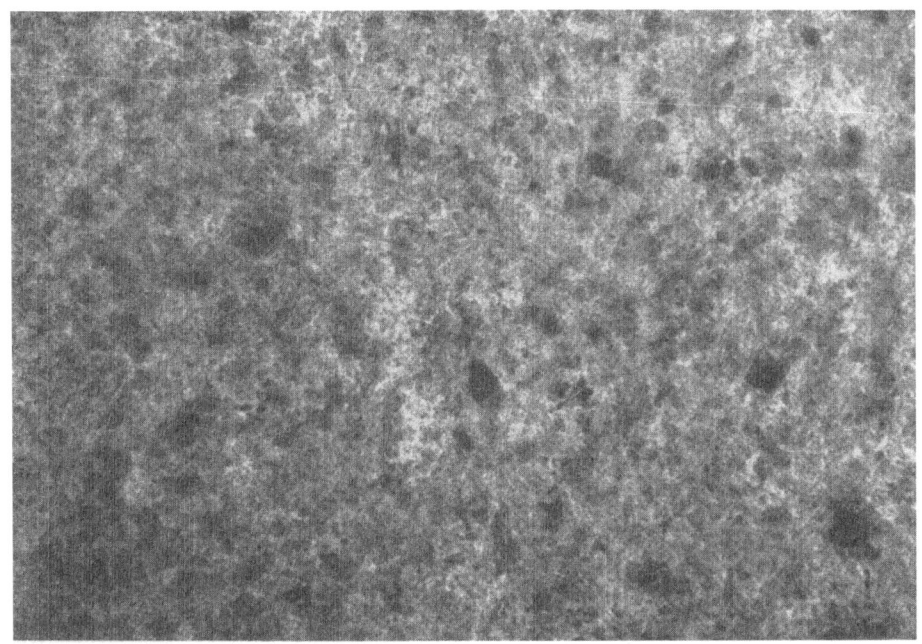

fig. 3. Art paper, magn. ca 130 X.

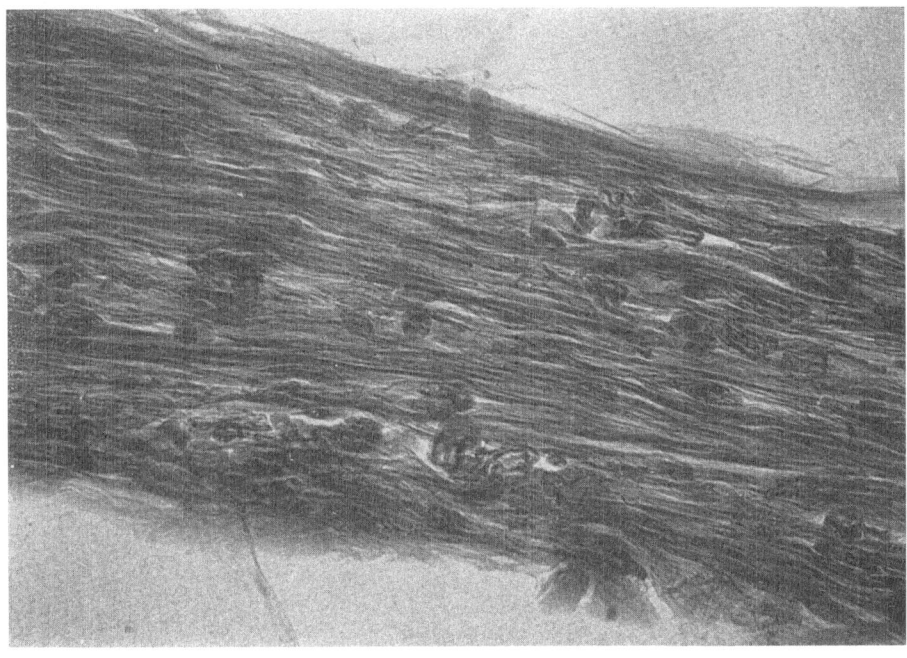

fig. 4. Transverse section of card paper, magn. ca 200 X.

fig. 5. Longitudinal section of (unknown) wood.

fig. 6. Inclusions in aluminium foil, magn. ca 320 X

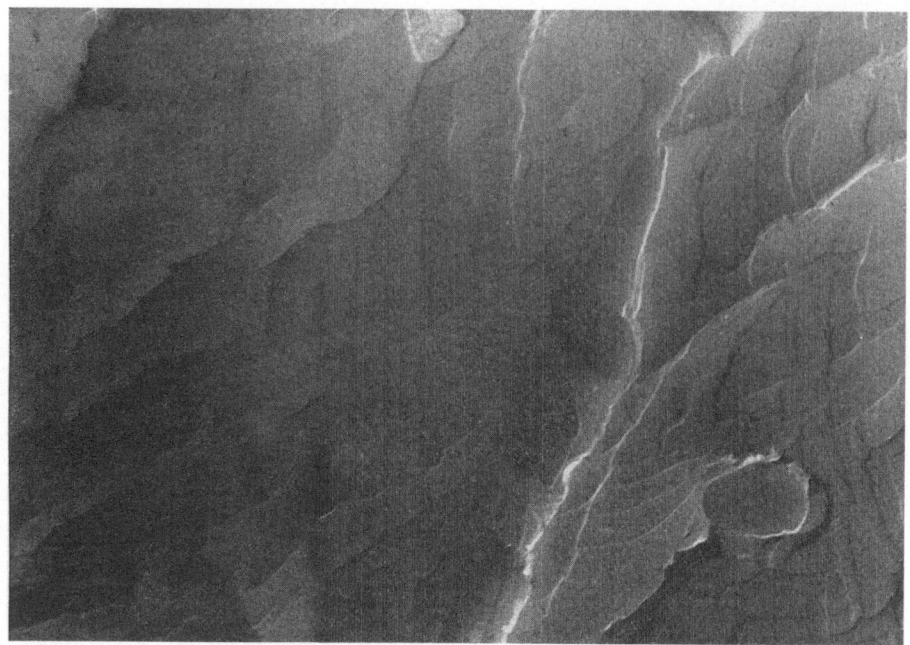

fig. 7.   Replica of a surface of fracture (Araldit), magn. ca 80 X

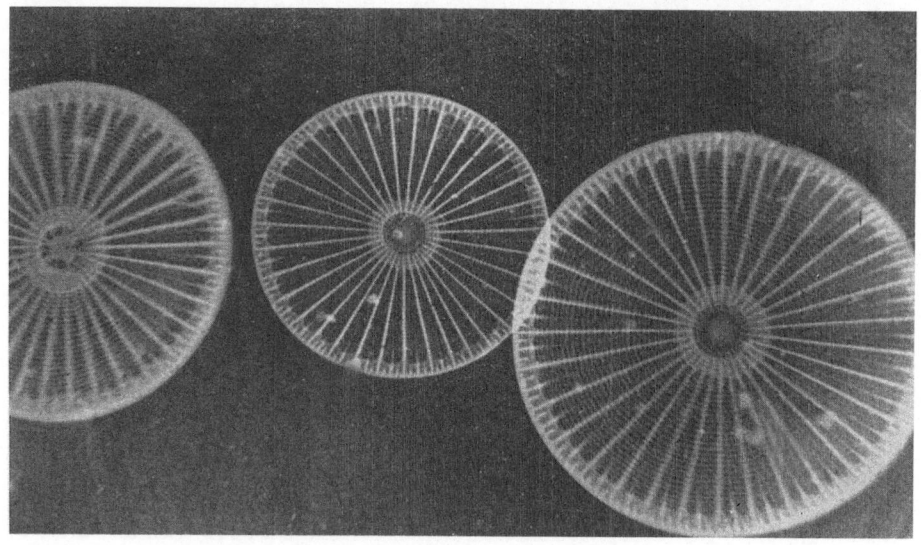

fig. 8.   Diatoms (Arachnoidiscus) gold shadowed, magn. ca 180 X

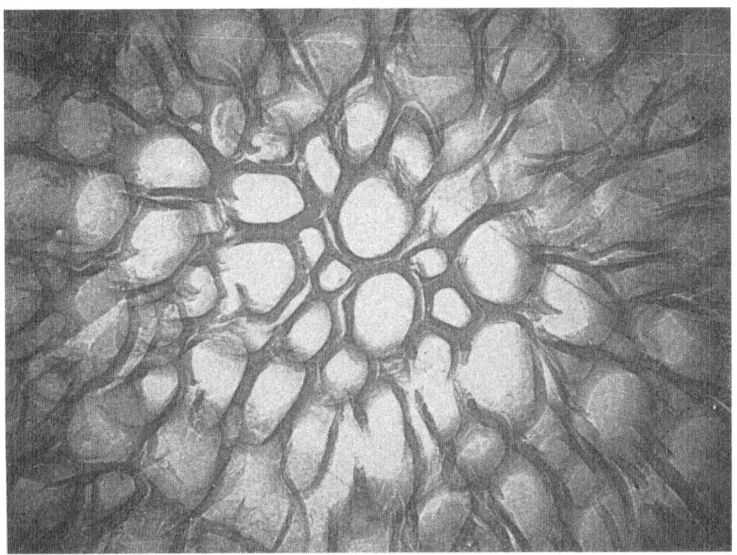

fig. 9. Transverse section of ashwood, magn. ca 400 X

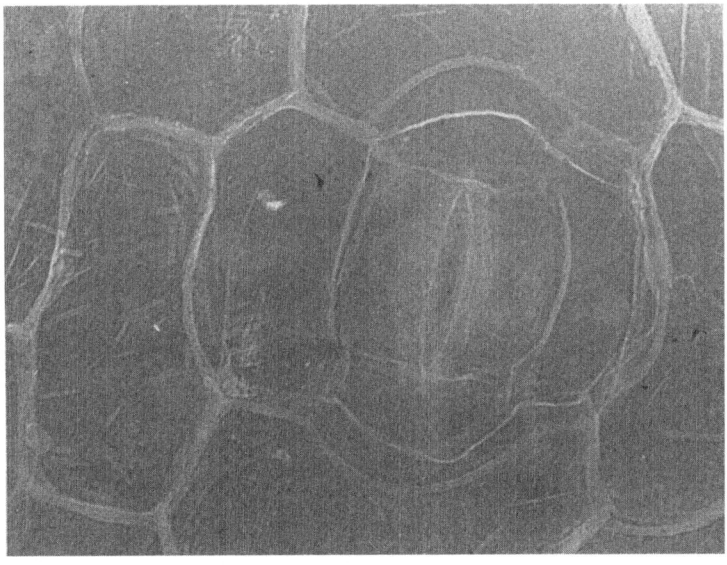

fig. 10. Stigma of a leaf, magn. ca 520 X. Recorded on ultrafine grain film,
initial magn. ca 4 X.

fig. 11.   Feather, magn. ca 65 X

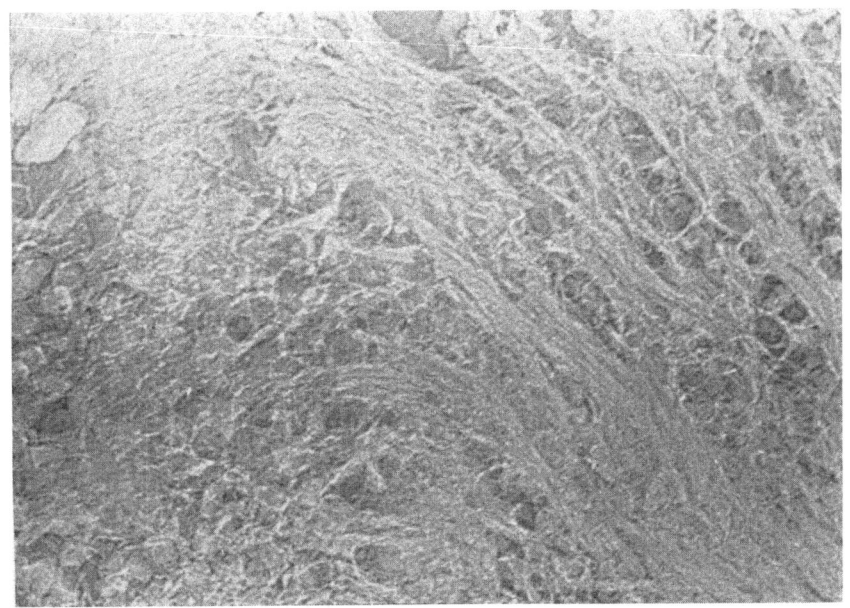

fig. 12. Human appendix, magn. ca. 520 X. Recorded on ultrafine
grain film, initial magn. ca 4 X

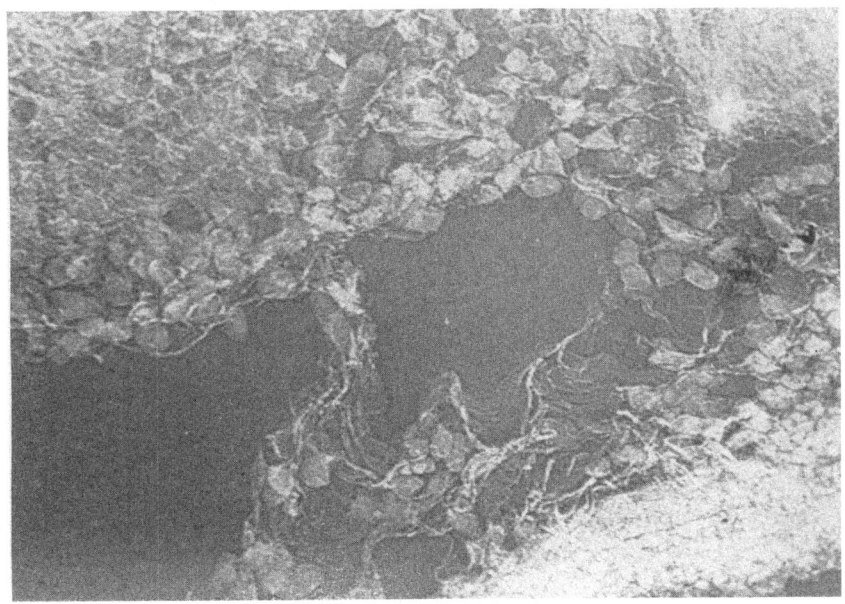

fig. 13. Human appendix, magn. ca 520 X. Recorded on ultrafine
grain film, initial magn. ca 4 X

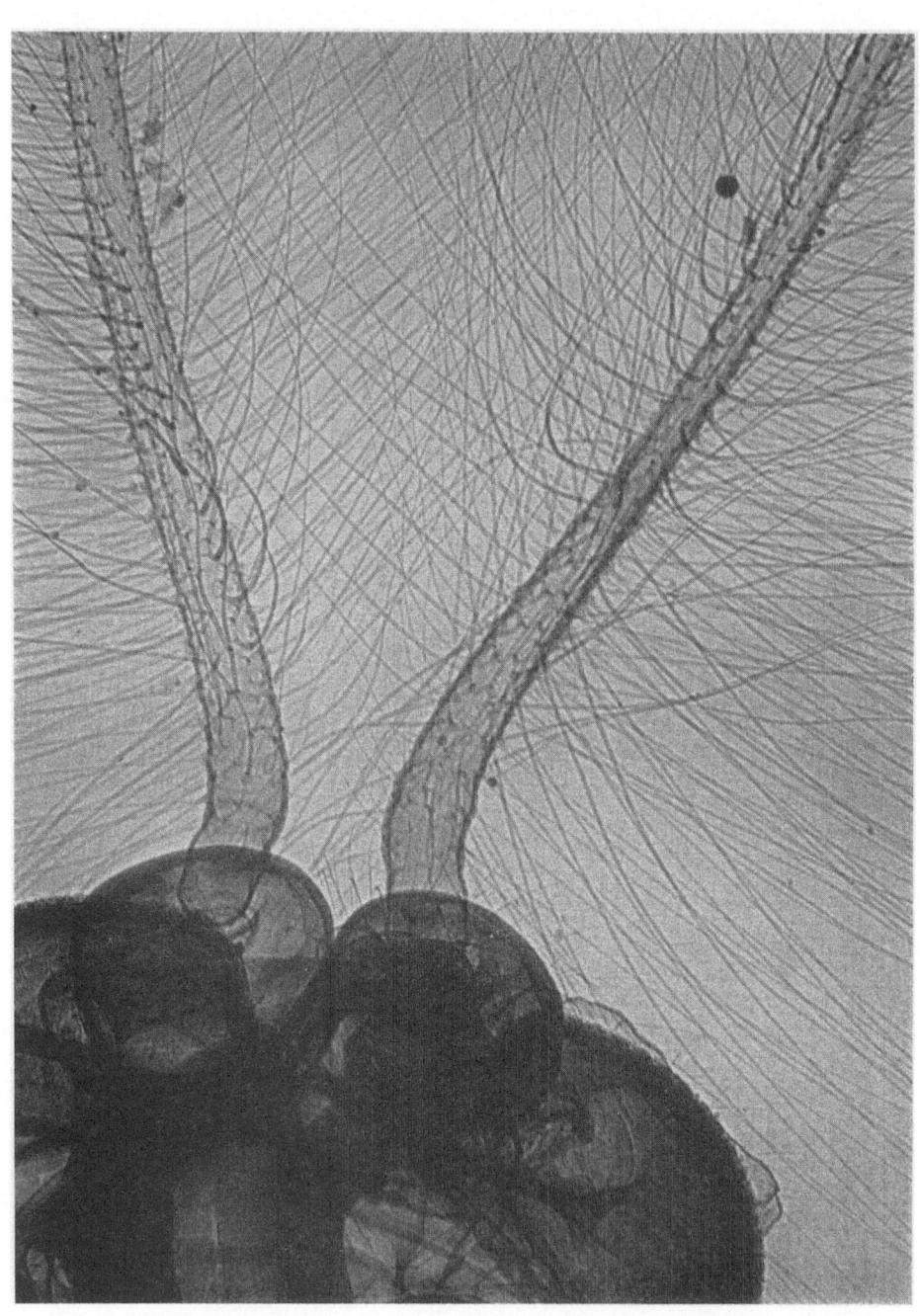

fig. 14. Head of a mosquito, magn. ca 140 X.

SUMMARY

As X-rays are hardly refracted in matter X-ray lenses have little practical importance for imaging purposes. Therefore the 5 existing X-ray microscope types, briefly described in chapter I, do not use X-ray lenses. The development of one of these types, the projection microscope, forms the main theme of this thesis.

Besides the requirements for a microscope in general, among other things a good resolving power, it is desirable that in the X-ray microscope the radiation gives a contrasty image. In chapter II the consequences of that fact are discussed. As a result of the limited specific emission of the hot cathode in combination with the aberrations of the electron lenses, the total radiation intensity appears to be very low. The maximum realizable image quality is now determined by the minimum radiation intensity which is necessary for visual focusing.

Chapter III deals with an improvement obtained by the development of a focusing method which works independently of the X-rays. This has been realized by using electrons which are elastically reflected at the target. They pass the lens in the opposite direction and give an enlarged image of the focus at the electron source. As the distance of this secondary image from the optical axis is a function of the centring position of the electron optical system it is possible, moreover, to center the apparatus very accurately.

Because the X-ray images are recorded photographically chapter IV deals more in detail with the properties of the film materials. It appears that the concepts sensitivity and gradation are of little interest with X-rays, and that the resolving power is not an essential film quantity. A definition for the film quality, applied to projection microscopy, is proposed. Furthermore the gradation adaptation when printing is examined more closely, and the properties of the fluorescent screen are compared with those of a film.

The ultrafine grain films as described in chapter V appear to have better properties than the normal fine grain films when using non-monochromatic rays. Besides the fact that they absorb less hard rays, there are reasons for assuming their quality to be higher; this is clearly demonstrated in fig. 5 of chapter V. A short reconsideration of the 2X method is given as a result of the further understanding obtained about the film properties.

In Delft the practical use of the projection microscope has been confined to morphological studies of objects. The calculations in chapter VI show that the wavelength of the radiation used must be such that $\mu z = {}^4/_3$ for maximum information. If the X ray dose must be kept to a minimum μz must be ≈ 2.2. Some preparation techniques are studied. The conditions for making good stereoscopic exposures are deduced, and the necessity for a calibrated object shift is shown.

Finally in chapter VII a short description is given of an X-ray microscope which is being built for the State University of Ghent. This apparatus is being constructed by the Technical Physics Department, T.N.O. and T.H., Delft.

SAMENVATTING

Doordat röntgenstralen in de materie vrijwel niet worden gebroken, hebben röntgenlenzen voor afbeeldingsdoeleinden weinig practische betekenis. De 5 bestaande röntgenmicroscooptypen, die in hoofdstuk I in het kort worden beschreven, maken dan ook geen gebruik van röntgenlenzen. De ontwikkeling van één van deze typen, nl. die van het projectiemicroscoop, vormt het hoofdthema van deze dissertatie.

Naast de eisen die we aan microscopen in het algemeen moeten stellen, o.a. een goed scheidend vermogen, wordt van het röntgenmicroscoop verder nog verlangd, dat de gebruikte straling een voldoend contrastrijk beeld geeft. In hoofdstuk II worden de consequenties daarvan besproken. Als gevolg van de beperkte specifieke emissie van de gloeikathode, gecombineerd met afbeeldingsfouten van de gebruikte electronenlenzen, blijkt de totale stralingsintensiteit zeer laag te zijn. De maximale beeldkwaliteit die realiseerbaar is, wordt nu bepaald door de stralingsintensiteit die minimaal nodig is om visueel te kunnen focuseren.

Hoofdstuk III handelt over een verbetering, die wordt verkregen door de ontwikkeling van een scherpstellingsmethode, die onafhankelijk van de röntgenstralen werkt. Dit is verwezenlijkt door gebruik te maken van de electronen die elastisch tegen de trefplaat worden gereflecteerd. Deze passeren de lens in tegengestelde richting en geven ter hoogte van de electronenbron een vergroot beeld van het focus. Daar de afstand van dit secundaire beeld tot de optische as een functie is van de centreringstoestand van het electronenoptische stelsel, is het bovendien mogelijk, het apparaat zeer nauwkeurig te centreren.

Daar de röntgenbeelden steeds langs fotografische weg worden vastgelegd wordt in hoofdstuk IV nader ingegaan op de eigenschappen van de filmmaterialen. Het blijkt, dat de begrippen gevoeligheid en gradatie voor röntgenstralen weinig interessant zijn en dat het scheidend vermogen geen essentiële filmgrootheid is. Een definitie van de filmkwaliteit, aangepast aan de projectiemicroscopie, wordt voorgesteld. Verder wordt nader ingegaan op de gradatieaanpassing bij het afdrukken, en worden de eigenschappen van het fluorescentiescherm met die van een film vergeleken.

De ultrafijnkorrelige filmsoorten blijken, zoals in hoofdstuk V beschreven wordt, bij gebruik van niet-monochromatische stralen betere eigenschappen te bezitten dan de normale fijnkorrelige. Behalve dat ze minder harde stralen absorberen zijn er redenen om aan te nemen, dat de kwaliteit hoger is. Dit wordt duidelijk gedemonstreerd door fig. 5 in hoofdstuk V. Een korte nabeschouwing van de tweemaal methode wordt gegeven in verband met de verkregen inzichten over de filmeigenschappen.

Het practisch gebruik van het projectiemicroscoop bleef in Delft beperkt tot morfologische studies van objecten. De berekeningen in hoofdstuk VI tonen aan, dat de golflengte van de gebruikte straling zodanig moet zijn, dat $\mu z = {}^4/_3$ voor maximale informatie. Indien de röntgendosis minimaal moet worden gehouden dient $\mu z \approx 2,2$ te zijn. Enige prepareertechnieken worden bestudeerd. De voorwaarden voor het maken van goede stereoscopische opnamen worden afgeleid en de noodzaak van een controleerbare objectverschuiving aangetoond.

In hoofdstuk VII wordt tenslotte een korte beschrijving gegeven van een röntgenmicroscoop, dat voor de Rijksuniversiteit te Gent wordt gebouwd. Dit apparaat wordt vervaardigd door de Technisch Physische Dienst, T.N.O. en T.H., Delft.